A Crowning Achievement

A Crowning

Achievement

130 YEARS OF

INNOVATION,

PERSEVERANCE

AND TRUST

We dedicate this book to our children, Ellen, Kathy and David.

International Standard Book Numbers
ISBN-10: 0-9816022-0-7
ISBN-13: 978-0-9816022-0-2

2008923218

This story of the Crown companies starts with two intrepid Swedish immigrants setting up business on the banks of the Mississippi River in the 19th Century. It continues today, 130 years later, with a company whose international scope and size would amaze its founders.

Our story is full of twists and turns, calculated risks, some sobering "near-death" experiences and limitless energy and perseverance. Our grandfather, Elias, and our father, Clifford, are among many people who steered Crown through 130 years.

Crown people are bright, dedicated, "can-do" folks who are adaptable, resourceful and hard-working. They are attracted to our company's unique culture where people truly respect each other, they take initiative *and* responsibility without being told to, they eschew company politics and they know how to have fun. Even after 130 years and many changes, we are still a family.

Crown is successful because of these people. They have created a company that is trustworthy, committed to providing the best quality and expertise in our field, open to inventive new ideas (wherever they come from) and dedicated to making and keeping strong, respectful relationships in business.

The people featured in this book are a representative sampling of many more people who contribute to the Crown story.

Only a fraction of one percent of all American companies reach 130 continuous years in business. We have—and our story tells you why and how we did it.

Cliff Anderson

George Anderson

Table of Contents

Opportunity, Grit and Craftsmanship

Historians called it a Swedish "surge." By the 1870s, record numbers of Swedes unpacked their bags in Minneapolis and Saint Paul. While their countrymen who preceded them sought farmland on Minnesota's rural frontier, this ambitious wave of opportunity-seekers picked the Twin Cities. Burgeoning lumber and flour mills promised jobs and prospects for entrepreneurial ventures in both cities. But for most Swedes, Minneapolis seemed a better bet. The Germans and Irish, after all, already dominated Saint Paul.  The "mill city" was a magnet for enterprising

(Left to Right): Crown employees gather at the Tyler Street plant in Northeast Minneapolis during the early 1900s. Crown produced iron work for many downtown Minneapolis landmarks, including the Foshay Tower. Patterns for ornamental iron were intricate. For its contributions during World Wars I and II, Crown was recognized with the U.S. government's "E" for excellence award.

Swedes and—before long—
70 percent of all Swedes in the
Twin Cities chose the younger twin
for their home. Three of those new
immigrants were Andrew Nelson,
E. Hernlund and August Malmsten.

Not long after arriving in
Minnesota, Nelson and Hernlund
opened a blacksmith shop in 1876
to sharpen saws for the mills, shoe
horses and craft tools for lumber-
jacks. They picked a strategic spot: a
small wooden building at 122 Main
Street Southeast on the banks of
the Mississippi River—close enough
to feel the spray of the St. Anthony
Falls. They were just a short walk from the multiplying mills that depended on the falls for
cheap, abundant power.

Crown Iron Works newspaper
advertisement, circa 1904

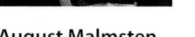

August Malmsten　　**John Hernlund**　　**Andrew Nelson**

"The two enterprising blacksmiths were skilled artisans in iron work. . . . handsome, strong men with unlimited energy and a compelling desire to make good in their new country."

Clifford H. Anderson in the Crown Iron Works Centennial book

Though only in his 40s, Hernlund died just two years into the new venture, leaving his
younger business partner, Nelson, age 29, alone. To generate extra income, Nelson rented
space in his little shop to August Malmsten, a railroad machinist. Within months, proximity led
to trust and the pair formed a new business partnership in 1878 called Malmsten, Nelson and
Company. "This was the real beginning of Crown Iron Works Company," Clifford H. Anderson
remembered decades later in his own historical account. "They began to manufacture machinery, as well as bolts and tools." In addition, thanks to Andrew Nelson's training in ornamental
iron, the venture offered this specialty, which was popular with architects and builders. When
the late E. Hernlund's son, John, chose to invest in the business and join the staff, Malmsten,
Nelson and Company also boasted its first university-trained mechanical engineer.

The little company with tall aspirations employed eight people and generated $8,000 in
1879, one year after the Washburn "A" Mill exploded, destroying half the milling district along the
west bank of the Mississippi. That calamity might have discouraged some, but not the Swedes.
They didn't give up even when their own shop was annihilated by fire in 1882. They simply
anted up the cash to buy a little stone building that had been St. Anthony Falls' original city hall
at 113 Second Avenue Southeast. Imbued with optimism, the blacksmiths-turned-iron-artisans
officially incorporated under their new name, Crown Iron Works Company, on January 6, 1884.

Tackling the "Tough Assignments"

Those early years of Crown Iron Works were a struggle. The country was still reeling from the
aftermath of the Civil War. Banks weren't betting on little start-up companies and there were
no reserves to fall back on at Crown. When money was short (as it often was), the company's
officers simply went without pay.

Organic Growth: the "Slow, Hard Way"

Many Swedes arriving in Minnesota settled along the Saint Paul and Pacific railroad line, stretching west of Minneapolis through Wright, Meeker and Kandiyohi counties, thanks to Hans Mattson, an immigration promoter and railroad land agent. There was abundant farm land and settlers wrote home to their friends in Sweden, urging them to make the trip. Dassel, Minnesota, was one of those towns and Elias (Eli) Anderson became one of its favorite sons.

> "Very little—if any—of the capital came from anybody outside the organizers themselves. The individuals put up the money as entrepreneurs who were willing to risk everything. . . . This is the way Crown Iron Works started."

Clifford H. Anderson

At 17 in 1886, Eli Anderson left his farm home and headed to Minneapolis, landing a job as a baker's boy. In his spare time, he attended public school and added courses from Archibald Business College. By 1889, the 20-year-old Anderson applied for—and landed—a $30-a-month job as bookkeeper at Crown Iron Works. Clean-cut and presentable in his wire-rimmed glasses and trimmed handlebar mustache, Anderson was eager and ambitious, a religious man with strong convictions. "He was determined to become the mainspring of Crown Iron from the very day he started as a bookkeeper, complete with a high desk, gas light and a green eyeshade," his son, Clifford Anderson, said.

When Eli joined Crown, the company had a foundry; it was crafting architectural metals, including cast-iron columns, and it produced structural steel for buildings in Minneapolis. In addition, the company sold the patented Blackmer Saw Sharpener and a "gummer" for setting and sharpening large circular saws. Crown was already showing an interest in attracting inventors with ideas worth patent protection.

Crown blacksmiths in their new Tyler Street foundry, circa 1912. Eli Anderson joined Crown in 1889 as a $30-a-month bookkeeper.

There was only one way to succeed in those days before venture capital. "It was by investing their work and skills to the limit and putting every dollar they could earn and save into the business," Eli's son observed. "Very little risk capital was available, so growth had to come from earnings, the slow, hard way. Only after a company was established as a proven and attractive investment were the banks interested in lending to the new, little organizations."

Crown's founders would soon discover that Eli Anderson had the grit and energy to carry the company through many a discouraging event. Years later, Eli remembered the financial panic of 1893: "It lasted until 1898 and all businesses suffered," he told a newspaper reporter, "but we hung on."

For his dedication and grace under economic fire, Eli Anderson would be elected President and General Manager of Crown Iron Works in January 1926.

During the first 30 years of the 20th Century, Crown's ornamental bronze and iron work graced (from left) the Minneapolis Star and Tribune building, Foshay Tower and Medical Arts Building in Minneapolis, Minnesota; First National Bank of Fort Worth, Texas; and theaters in Detroit, Michigan.

> "Crown had a big part in the building of Minneapolis. The company was becoming known as a leader in producing ornamental iron for the whole Midwest, from Duluth to Iowa, and into the Dakotas."
>
> **Clifford H. Anderson**

Luckily, a few big jobs eased the financial strain. When construction began on the Guaranty Loan Building (later called the Metropolitan Building) in downtown Minneapolis, the 12-story design was the first "skyscraper" in the region. Crown produced all the intricate cast- and wrought-iron work that formed the building's interior court, which rose through all 12 stories. The two-year job, started in 1888, was a godsend. On top of that, Crown was tapped to produce stairs and railings of cast iron and bronze, along with supporting beams and columns for the Minneapolis City Hall and Courthouse, a six-year project that began in 1889. Only a few blocks away, Crown also contributed to construction of the ornate Federal Court House on Marquette Avenue at Third Street, built around 1900.

These jobs were weighty and required thousands of man-hours: "Imagine, if you will," Clifford Anderson wrote, "the cutting of steel beams with cold chisels, handling them in an open yard with a simple, hand-operated crane and the transportation of many heavy pieces with horses and wagons.

"Try to visualize the amount of forge work required to produce an elevator enclosure or an ornate iron and bronze railing, especially when hundreds of these were required.

"And then try to conceive of casting the many ornate 20-inch fluted and ornamental columns for the Metropolitan Building, some 20 feet high. Few foundries would dare tackle such a tough assignment."

But Crown did, willingly and often, during those formative years. In fact, Crown's plant on the Mississippi River was getting cramped. When the portion that housed Crown's foundry burned in 1905, the company's officers took a big step toward progress: they bought property on the outskirts of Minneapolis at Thirteenth and Tyler Streets Northeast; built a new foundry, machine shop,

structural fabricating shop and offices; and moved in two years later. By that time, the company's annual net earnings totaled $11,665 and its net worth—thanks to the new location—topped $75,485. But the decision to expand was not one that August Malmsten could abide. He exited Crown in 1907 and started a competing company at Crown's old location, calling it Atlas Iron Works.

Every Kind of Building Imaginable

During the first 30 years of the twentieth century, Crown's list of ornamental and iron work contracts multiplied and introduced the Crown name all over the United States. In the Twin Cities, the company's work graced buildings and a stadium on the University of Minnesota campus; in Minneapolis, the Builders Exchange, Roanoke, Medical Arts and Baker buildings, Foshay Tower, Albrecht Furs, Hennepin Avenue United Methodist Church, Central Lutheran Church, Calhoun Beach Club, The Woman's Club of Minneapolis, and the Radio City Theater building; the 100 First Avenue Building in Rochester, Minnesota; the Sioux City, Iowa, courthouse and Valley National Bank in Des Moines, Iowa; the Orpheum Theater and the St. Louis County jail in Duluth, Minnesota; Hibbing (Minnesota) High School; Republic Bank in Dallas, Texas; the Film Exchange, Maccabees and Francis Palms buildings, the Fisher Theater and the Fox Theater, all in Detroit, Michigan; the Cincinnati Times Star newspaper building and Coca-Cola Bottling Company in Cincinnati, Ohio; the Mutual Home and Savings Building in Dayton, Ohio; Boston and Springfield, Massachusetts, post offices; and the Agricultural Extension Building in Washington, D.C.

Other ornamental and iron work projects followed over the decades, until the cost and intricacy of this work lost favor among architects and builders: St. Mary's Hospital in Rochester;

> "A hundred years ago, iron railings, balconies, grilles and roof ornaments were considered integral parts of buildings, without which they would look unfinished—like a suit without a tie."
>
> **Linda Mack, architecture reporter for the *Minneapolis Star and Tribune*, November 9, 1986**

in Minneapolis, the Minneapolis Star and Tribune Building, Farmers and Mechanics Savings Bank, North American Life and Casualty, the Lutheran Brotherhood, Prudential, and American Hardware Mutual buildings, the First Unitarian Society, Donaldson's Department Store and the Minneapolis and St. Louis Railway Company; in Saint Paul, the Degree of Honor Building and 3M's headquarters in an adjacent suburb.

Crown had competition during its long run in ornamental iron work. In fact, its best rival, Flour City Ornamental Iron Company, hailed from Minneapolis, too.

Steel Posts by the Millions

Three years passed between the assassination of Austrian Archduke Francis Ferdinand in 1914 and the time when America's first armed division arrived in France to join World War I. Back in the states, Crown joined the war effort by manufacturing more than 2,000 tons of riveted steel girders for cargo vessel components and inventing a better way to make barbed-wire "entanglement" fences used to protect soldiers in the trenches. H. Nicoud, a staffer in Crown's engineering department, designed a machine to make these steel screw posts that held the

> "At Crown Iron, skilled craftsmen have learned that the beauty of ornamental work is more than pure dimensions. Highly adept workers etch cutting marks, saw, polish and examine each piece of metal with trained eyes that let no flaw go by."
>
> **Early Crown Iron Works advertising copy**

A Fur Piece

It was one of the more unusual ventures, but Crown Iron Works was always nimble and open to opportunities. It started with fur farmers.

During the 1920s, fur coats and wraps were a popular fashion statement. Who can forget the muskrat coats donned by dapper college men and beaver capes encircling the shoulders of female fashionistas called "flappers"?

Several Minnesota fur farmers approached Crown, asking them to make specially designed fences to pen their foxes, beavers, muskrats and other animals. Unwilling to overlook an opportunity, Crown bought fencing, posts, and special cage wire and sold these assembled products to farmers all over the United States, even Alaska and as far away as Finland. "Noting the demand," the Minneapolis Star Journal *observed in 1928, "Crown established a regular fur fence department with the result that it now builds not only fences, but other equipment needed by the breeder of fur-bearing animals." In about five years' time, Crown built that unexpected venture to an estimated $450,000 in sales in 1928, alone.*

Seventy-five years later in the dusty Crown Iron Works store room, researchers would uncover a little-known set of files, curiously named "Fences and Fur."

barbed wire. The company produced millions for the global conflict and, again, in World War II, three decades later.

With the War's end in November 1918, Crown returned to its peacetime ornamental iron work, budding structural steel business and a couple new ventures: manufacturing and selling Nilson tractors and building fences—all kinds of enclosures prompted, no doubt, by its wartime experience with screw posts. Crown celebrated its first structural steel contract in 1916 when it landed the St. James Hotel project in Minneapolis, one of the city's early steel frame buildings. Their next big project would be Marshall High School 12 years later.

Remarkably, the company's net worth had multiplied nearly five times in 15 years, to $373,396 by 1920. And by 1928—the company's 50th anniversary—the company's revenues reportedly reached $1.5 million.

The Roar Before the Rout

The "Roaring Twenties" was no overstatement. This was the era of unprecedented business innovation and stock speculation. During these giddy boom years, Sosthenes Behn started ITT, F. L. Maytag introduced his washing machines and Jacob Schick patented the first electric razor. Ida Rosenthal debuted the Maidenform bra and Tom Watson left National Cash Register to start IBM.

New companies flourished like spring wheat and the marketplace was dazzled. In 1925, the Dow Jones Industrial Average exceeded 100 for the first time in history and Americans who never dared invest before put down their money. Capital was the fuel of the nation's industrial expansion and by 1928, the Dow broke 300 and then hit the high-water mark—381—on October 24, 1929.

They called it "black Thursday" when the Dow dropped an unprecedented 21 points that day and the trading volume spiked to nearly 13 million. The awful descent continued and five days later, on October 29, Big Board companies had lost a total of $30 billion in market value. (In 1932, the market bottomed out at only 41 points.)

Stockholders were wiped out. Within months, hundreds of thousands of U.S. workers were laid off. For four years afterward, an estimated 100,000 people lost their jobs each week and most turned to free soup lines. Even today, we call that ten-year era the Great Depression.

"The stock market crashed in 1929 and money just disappeared," Clifford Anderson remembered. "We estimated a few projects and landed them in 1930 and 1931, but these resulted in heavy losses. When we completed them in the early 1930s, they made a sorry mess of Crown's finances."

No one dared think the Depression would last ten years, but it did. There was nothing being built, save for a few government buildings. Subcontracts were auctioned off by contractors at prices far below cost. "We followed a policy of taking work sufficient to pay our key employees," said Anderson, "with the intention that we cut our expenses to a bare subsistence level in order to survive."

Crown's board minutes were cryptic in their reporting: 1930 "Big loss this year." 1931 "There was a big loss this year." 1932 "Another bad year." 1933 "A smaller loss this year." 1934 "Another loss." 1935 "In the red again." The repetition was heartbreaking.

A U.S. Army illustration of barbed wire with screw posts, manufactured by Crown, and used to secure perimeters during World Wars I and II.

Crown made and sold Nilson tractors after World War I.

"The Depression years were debilitating, losing years—a period that nearly terminated our history in a sad ending."
Clifford H. Anderson

(Upper Left): Crown produced 20,000 tons of metal decking and fittings for portable "pontoon bridges" used to move men, trucks and tanks across water and impassable terrain during World War II. (Upper Right): Crown fabricated sections of this U.S. Navy oil tanker launched by Cargill's Savage, Minnesota, shipyard. The railings on the Mendota Bridge above the ship were also made by Crown.

Trust in the Lord— but keep two gages of water on your own account.

This quotation hung for decades in Crown's boiler room and remains today in the Chief Engineer's office.

Crown cut salaries in half and took paying work, no matter how little it generated. "Management was forced to make a very difficult decision," Clifford Anderson said. "Shall we take work at competitive (devastating) prices and cut our costs accordingly or shall we take the other track and just close up shop and wait out the Depression?"

Crown's leaders chose the former, believing that if they shut down, their employees and families would go hungry. Some of the company's officers borrowed against their life insurance and went without regular compensation for more than two years. At one point, Clifford Anderson had less in his bank account than one of Crown's pattern makers, who floated his boss a little loan.

In the midst of those tough years, the AFL Union organized Crown's workers in 1935 and orchestrated a 13-week strike. "We became a union shop," Anderson recalled. "As a result, all of our wages went up by 20 percent and our business dropped still more. Seemingly, there was no hope."

The company turned to making more screw posts (à la World War I), experimented with manufacturing new types of conveyer screws of all sizes and invented popular coal "stokers" for home furnaces and commercial buildings. Crown helped build a jail for the City of San Diego and a dredging machine for California's gold fields. The company squeaked by.

Standing Tall in WWII

World War II would become the world's largest, costliest and deadliest war in history. By the summer of 1940, when Adolph Hitler had completed his brutal occupation of Poland and 188 Nazi divisions swarmed through Russia's Ukraine region, America issued a call for large-scale

war production. President Franklin Roosevelt's declaration of neutrality in 1939 could not stand for long and Japan's bombing of Pearl Harbor on December 7, 1941, made America's entry into the war official.

The U.S. Marine Corps ordered 50,000 barbed-wire entanglement screw posts and Crown's superintendent, Andrew Anderson, went to work improving the manufacturing methods and equipment Crown had developed during World War I to produce them. He designed winding, looping and shearing machines that increased the company's output—just in time for more orders. In all, Crown produced 20,000 tons of screw posts—an estimated seven million. They were transported to the Aleutian Islands, Midway, Hawaii, New Guinea, Australia and Africa, and they were used along both of America's coastlines for defense.

But that wasn't all. Crown produced 20,000 tons of metal decking and fittings for portable "pontoon bridges" that the Allies used to move men, trucks and tanks across water and impassable terrain. "When American Army engineers throw a span across a river, stream or ravine on the way to Berlin or Tokyo, chances are good a Minneapolis war plant had an important role in building that structure," the *Minneapolis Sunday Tribune* declared on January 9, 1944. "Such was the case when Allied forces crossed the Volturno River in Italy on a pontoon bridge. The pontoon bridge, it can now be revealed, is one of approximately 200 types of war equipment manufactured, either in part or entirely, by the Crown Iron Works Co." *The Tribune* reported that Crown also made portable steel highway bridges, many sent to India for use on vital supply roads.

Crown used its structural steel shop to fabricate ship sections for oceangoing tankers and Army towboats, supplying Walter Butler Shipbuilding Company in Superior, Wisconsin, and

Crown employees answered the call during World War II, producing frame ship sections (background) and screw posts (foreground) at the company's Tyler Street plant.

Getting Their Act Together

It was an unexpected union of aspiration and geography. Clifford Hawkins Anderson came from a family of Minnesota Swedes. His grandfathers immigrated to America, finding work as farmers and nurserymen on the Midwest plains. His father, Eli—the adventurous one—ventured to Minneapolis while still in his teens to find education and opportunity.

Katharine Irving came from a successful family of stage and movie actors who moved from New York to Hollywood in the 1920s, when America's infant movie industry settled on the West Coast.

The couple met on a blind date. Young Katharine had acting aspirations and experience, too, but her attraction to Clifford Anderson was stronger. Earning his civil engineering degree from the University of Minnesota and pursuing post-graduate study in business at Stanford University (he received Stanford's first MBA diploma in 1928), Anderson was an outdoor-loving jock. He was a scholar who crewed on the winning Iron Wedge rowing team and managed the University of Minnesota football team, while earning membership in the honor society. He loved duck hunting in his native state. As a teenager, young Clifford Anderson joined the Crown Iron Works payroll doing odd jobs for 25 cents an hour.

While his father, Eli, joked with his son to keep him humble (on his first payroll check, Eli wrote, "I doubt if he's worth it…"), the boy had the qualities and qualifications to succeed his father.

Right out of Standford, Clifford Anderson worked for another steel fabricating company on the West Coast.

When Clifford and Katharine married in 1931, they returned to Minnesota and Clifford came back to Crown. Later they lived in a house Clifford designed. He remodeled his parents' garage in Southeast Minneapolis using remnants he salvaged from several 1920s mansions that were torn down during the Depression.

Clifford Anderson progressed through Crown's engineering department to the general superintendent's position. When Eli died in 1946, he became Crown's chairman and president.

Clifford H. Anderson would become known as a solid businessman, respected for the integrity of his decisions, his perseverance, civic commitment and appreciation of a balanced life. "One paneled wall of the office of Clifford H. Anderson, president of Crown Iron Works Co., is plastered with autographed photos of high Army brass, most of them in duck hunting garb," the Minneapolis Star "Town Toppers" columnist wrote in the 1940s. "The opposite wall has niches for products turned out by Crown, such as eagles over doors of Ford's first Dearborn plant. Anderson is right in the middle, figuratively as well as literally. He isn't sure whether he'd rather be shooting canvasbacks with generals of the Army Air Force or shooting the breeze with old-timers among the 250 employees of his metal fabricating company at 1229 Northeast Tyler Street. He does each equally well."

Clifford Anderson served on many nonprofit community boards. He was also quite active on the IDS Funds Board of Directors that included, as its members, U.S. President Richard Nixon and his Secretary of Defense, Melvin Laird.

In the 1950s Clifford Anderson produced successful fundraising "telethons" for the United Cerebral Palsy Association, bringing in several professional entertainers of the day. He was an unabashed advocate for the Minnesota work ethic: "We can do more with, for and because of our fine, home-grown Minneapolis craftsmen," Anderson was quoted in a promotional piece for the city.

But the hard-working Anderson wasn't averse to restorative relaxation. His pet peeve? The Minneapolis Star columnist described it this way: "The yowl that goes up in Washington every time a top Army man sneaks away from his desk for a few days of relaxation in the duck pass."

Anderson ought to know. After inspecting the output at Crown Iron Works, those uniformed emissaries from Washington joined Anderson on hunting trips to the Dakotas and, decades later, were guests at Anderson's log cabin on Little Whitefish Lake near Garrison, Minnesota.

Clifford H. Anderson (middle) shared his love of hunting with Crown salesman Cliff Stephens. When he applied for work at Crown (right), Cliff Anderson's father, Eli, joked that an hourly rate of 25 cents might be too high for his son.

Cargill, Inc., in Savage, Minnesota. Crown's machine shop made components for anti-aircraft gun mounts designed and manufactured by Northern Ordnance Company in nearby Fridley, and Crown molded castings for portable electric generating equipment made by another Minnesota business, Electric Machinery Company. On top of that, Crown turned over old equipment to be recycled into scrap metal and munitions.

A Near-Death Experience

While wartime orders breathed economic life back into Crown, production logistics could be punishing. In one case, the pontoon bridges called for wood sections, as well as aluminum and steel. But Crown had no expertise in wood fabrication. They had to solve the problem—fast.

Fortunately, a millwork factory stood idle close by. Crown contacted the owner of the bankrupt company, Albert Carlson, and asked him to join the effort. "All in all, Albert's company was to produce, for Crown alone, a total of over 1,200 carloads of wood parts during the war," Clifford Anderson said. "Crown Iron started the company up, financed its payroll for a time and contracted for materials."

Crown signed a big contract with the federal government containing a crucial proviso: missing the delivery date would cost Crown $100,000.

To locate enough clear, kiln-dried Douglas fir, Crown hired G. M. Stewart Lumber Company. The company's owner personally traveled west to secure the order and returned ten days later, downcast and empty-handed.

If Crown didn't deliver on its contract, it faced certain bankruptcy. Eli Anderson tapped his son, Clifford, to journey east to Washington, D.C. Young Anderson feared the worst and the government's contracting officer, Capt. John Seybold, sneered: "I see that your company doesn't know its business." No, Seybold scowled, Crown could not be released from its contract. Yes, the penalty charge would be levied.

"This was a severe jolt," Anderson remembered. "Under the circumstances, it seemed inevitable that we would fail because our efforts to find a source were fruitless." But Anderson demonstrated a persistence that characterized Crown for decades. He and his Crown colleagues placed calls, sent telegrams and hunted relentlessly until they found a contact with the West Coast Lumberman's Association, who gave them hope of a timber supply in Tacoma. "It took several hours of long-distance negotiations, but we found a source, and just in time," Anderson said. Crown had its lumber and the Navy had its pontoon bridges. The new timber contract even saved the government more than a million dollars.

"We were established as a recognized source on the approved list of the War Department. Without this, we would have been finished as a supplier," Anderson said. Being on that list was no guarantee of work, however. Crown still had to be the lowest bidder to win a project.

Clifford Anderson was proud of Crown's performance during World War II. "Even though we were low bidder against tough competitors and with no financial assistance, subsidy or machinery furnished by the government," he said, "we made a substantial profit because of our ingenuity and efficient execution."

Too much profit, as it turned out. Crown Iron Works and every American company involved

> "It was a close call—a lesson in Government contracts we did not forget."
>
> **Clifford H. Anderson**

Crown received three Army-Navy "E" awards during World War II. Col. Lynn C. Barnes congratulates Eli Anderson while his son, Clifford, and employees look on.

in the war effort was subjected to the government's heavy-handed Renegotiation Board, set up to ensure against war profiteering. While there were some companies that took unfair advantage of the abundant wartime contracts, many others did not. But all were painted with the same brush. Crown had to turn back half of its profits one year and—of the profits that remained—Crown was taxed at 85 percent in 1943, 1944 and 1945. "There was little left to show for our efforts during World War II, except for the satisfaction of a job well done," Clifford Anderson concluded.

Recognition came to Crown when it was the first Twin Cities war plant to win the Army-Navy "E" (for excellence) Production Award in January 1944 (the company would receive two more). The occasion was celebrated with a U.S. Army color guard, music by the Marshall High School band, remarks by the Minneapolis Mayor and plenty of high-ranking men in uniform.

But it was Eli Anderson's thank you to Crown employees that carried the most weight. The reticent Swede said it was time to "say things that are in our hearts which we otherwise foolishly avoid expressing."

"We at Crown have been efficient in the production of materials of war for only one reason . . . that every man and woman has exerted himself in the execution of his particular task," Anderson said. "This has not always been easy, for there have been interruptions and sometimes seemingly insurmountable obstacles. . . . We all deserve to be proud, and can turn to the many men and boys of our families and friends who are at the fighting front with a feeling that we are doing our best."

To his longest-serving colleagues, Anderson gave special thanks: "I pay a special tribute to our employees who probably would otherwise be retired long since but for a feeling of duty toward our boys at the front," Anderson said. "These men, some with thirty to forty years of experience, have acted as lead men and teachers of younger and less experienced workers. They have done a great job in promoting a spirit of harmony and higher production."

It was this kind of longevity and loyalty that would ensure Crown's unfolding future. 👑

Re-Engineer...
or Disappear

When World War II officially ended with Japan's surrender on August 14, 1945, Crown Iron Works was in for a shock. Just eight days earlier, the United States had dropped an atomic bomb on Hiroshima, Japan, followed by another aimed at Nagasaki. Towering waves of change rolled toward America. ♛ With peacetime at hand, the U.S. government quickly dismantled its elaborate defense machine. "The first day after the end of the war, a full basket of telegrams arrived," Clifford H. Anderson remembered. "They all read the same: 'With respect to

(Left to Right): When World War II ended in 1945, Crown had to shift gears fast. Crown employees enjoyed after-hours social events, including bowling in the 1950s. When his father, Eli, died in 1946, Clifford H. Anderson succeeded him as Crown's President and General Manager. Crown embarked upon a whole new business when it gained the right to license a new soybean extraction process invented at Iowa State College.

"Uncle Sam has doffed his steel helmet, put on an apron, tucked a pencil behind his ear and assumed the role of store-keeper to conduct in Minneapolis, and elsewhere, the biggest sale in all history."

Minneapolis Star-Journal, August 28, 1945

War Department Contract No. 000, you are hereby ordered to cease all work hereon. Any costs incurred after this date will be at your own expense. You will be reimbursed only for labor and material costs already incurred unless further orders and authorization from this office are issued."

Many Minnesota companies, including Crown, had contributed mightily to the war effort—Honeywell, Cargill, Globe Shipbuilding, Minneapolis-Moline and General Mills, to name a few. Some, including Crown, had completely converted their normal peacetime operations to war production and there was no time to readjust. Prewar customers had disappeared and—literally overnight—companies were left with surplus raw materials from the war effort and nothing to do with them.

Members of the U.S. Corps of Engineers traveled with Clifford Anderson to the West Coast to inspect warehouses full of lumber intended for floating pontoon bridges. They poked through piles of steel, scrap and unfinished wartime products at Crown's plant in Minneapolis.

The Corps sold it all and reimbursed Crown with surprising speed. Crown was left to re-group. During the war, the pace was so intense that few could plan for the future. "Nothing had been allowed for the vast readjustment to normal peacetime operations," Anderson remembered. "Our whole plant had been rearranged to accommodate the manufacture of ship sections, barbed-wire entanglement posts, pontoon bridge parts, and many other items of special military nature." Crown quickly began rearranging its plant so that it could once again produce beams and columns, trusses, ornamental iron work and foundry projects. It was an expensive proposition.

But worst of all was Crown's lack of business. The company had lost contact with its regular customers and had to win them back. Equally tough, there were severe material shortages, especially steel. Crown was held to the same small quantity it bought during World War II, even though its prewar steel purchases were much larger.

Anderson even traveled to Pennsylvania's Bethlehem Steel Company, hoping to make a personal appeal. Standing in the Bethlehem office, waiting to meet a company officer, Crown's leader overheard the man instruct his secretary, "Tell Mr. Anderson I don't wish to see him." And that was that.

Just to stay in business, Crown bought some steel at retail warehouse prices. In 1946, its first year after the war's end, the company lost $33,800 with little chance of black ink anytime soon.

Inspiration From Henry Ford

In February 1946, Eli Anderson, age 76, died and his 42-year-old son Clifford succeeded him. In his desk, Eli kept files with new ideas for Crown products. In one file marked "Post War (confidential) Mr. E. L. Anderson Pres.," there was a description and engineer's drawing created by Crown employee Ken Mousseau. The proposed product combined the features of a rotary snow plow and a power mower that could revolutionize the industry, Mousseau said: "By the combination of snow plow, power mower and accessories, we eliminate the seasonal merchandise handicap and step into a year-round field, country-wide in scope." Good for homeowner and commercial applications, Mousseau believed the idea would take off with America's mass migration to the suburbs.

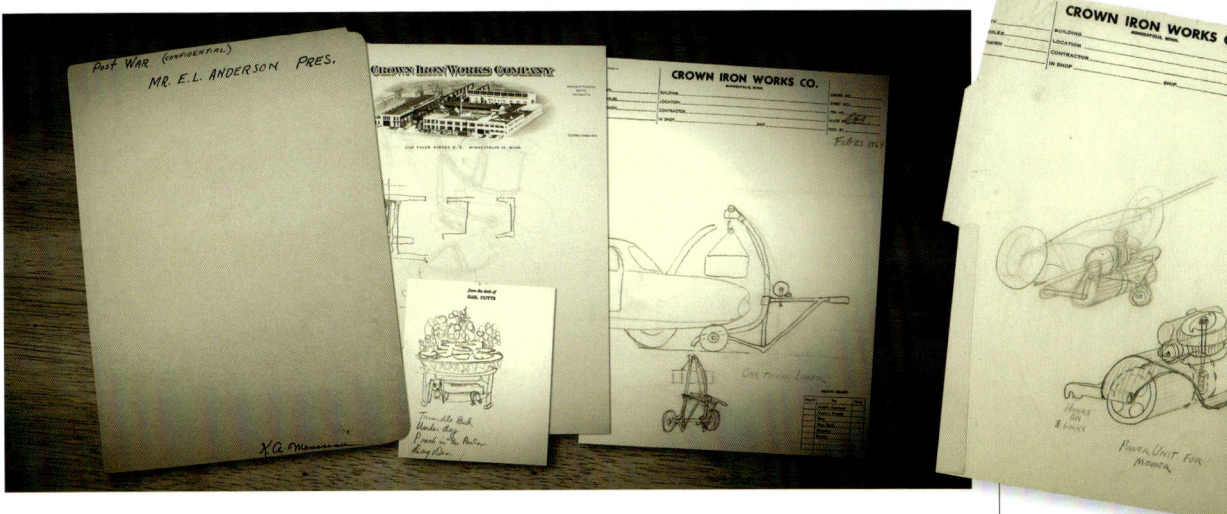

Hoping to replace revenues it lost after World War II, Crown investigated myriad ideas and inventions—some practical, some fanciful. Eli and his son, Clifford, kept them in a confidential file.

In other files describing possible new products that may have been worthy of a U.S. patent, there were descriptions of improved screw conveyors, better screw post machines, a meat chopper and a simple hoist. As the years passed, other inventions made their way into Crown's files simply marked "ideas" or "patents applied for"—a clamp for locking dock sections together, an improved chimney cap, a wire stretcher, a car trunk loader on wheels, a safety jacket, a tunneling device, an improved garbage can container, a utility cart that looked like a high-class wheelbarrow, and even a car-washing system. Several ideas made it into the market—the dock brackets, one-wheel trailers, car-top carriers for Sears Roebuck, the Crown sand mixer for foundry sand, and a machine that Crown manufactured on contract for years called the Porter Partition Assembler. Eli and Clifford Anderson were hungry for new products that could propel Crown into the postwar era and, as luck would have it, Americans were willing to invest and spend as never before.

But the best idea came to Crown from automaker Henry Ford.

As early as 1930, Ford started experimenting with soybeans at the Edison Institute in Dearborn, Michigan. "Ford had a lifelong conviction that industry must turn to the soil for many of its materials," Mildred Lager, author of *The Useful Soybean,* wrote in 1945. If America's farmers could produce commercially profitable products, Ford thought, the nation's economy could flourish (and more people would buy cars). Ford experimented with soybeans, starting with a small growing plot behind his own research labs at Ford Motor Company and expanding to huge farms yielding nearly 83,000 bushels. He put his crops through specially-devised oil extractors and explored what to do with the end-products.

Besides the high-protein content of soy for human and animal consumption, Ford imagined building an all-plastic car of the future by putting soy oil in baked-enamel finishes, paints, plastics molded to make accelerator pedals, coil covers and other car parts. Ford produced synthetic wool made with soybean yarn for car upholstery and others tried this wool for suits, topcoats and carpeting. Ford even used soy byproducts to make soaps and glycerin.

By the time America joined World War II, soybeans were proven emergency food rations.

"Ford perceived farmers as prime customers for his Model T automobile. If he wanted farmers for customers, he would have to find a way for industry to become a customer of farmers."

Journal of the American Oil Chemists' Society, **March 1977**

THE MINNEAPOLIS ST

MINNEAPOLIS, MINN., TUESDAY, NOVEMBER 9, 1948

BLAST DANGER CUT

New Process Takes Soy Mill to Farm Area

By HERB PAUL
Minneapolis Star Business Editor

A MINNEAPOLIS FIRM has been given exclusive rights for manufacture of a prefabricated soybean extraction plant using a non-explosive non-inflammable solvent which may revolutionize farm practices.

The plant is designed to be set up in the heart of the soybean producing area, like the grist mill of pioneer days, and return oil-meal feed to the farmer at a big saving in freight costs.

This was revealed today when Clifford Anderson, president of

(Upper Right): Iowa banker Lee Luick (left) introduced Crown to a new system for extracting oil from soybeans and Al Kaiser (right) helped design and build Crown's first pilot plant and process machinery. For a publicity shot, Luick and Kaiser ventured into a huge soybean bin.

"The farthest cry from logchain forging and other simple smithy work of Crown's early days is its soy bean project. . . . Crown Iron Works now builds a plant with 25 tons-per-day capacity which can be installed at any country elevator . . ."

Industrial Supply Expeditor, April 1949

A Far Cry From "Forging"

Clifford Anderson had been following Henry Ford's experiments with soybeans in the popular press, so he was receptive when Lee Luick, a retired Iowa banker, called on him. "Lee came to us one day in 1947 with a new product," Anderson said, "a system for extracting oil from soybeans and other seeds to make an edible animal food. A new method had been developed and experimentally proven by Iowa State College."

This solvent extraction research had started with Dr. Quincy Ayres, financed by Iowa State's development fund. Ayres' work was expanded and refined by Dr. O. R. Sweeney, long-time director of Iowa State's chemical engineering department, and Dr. L. K. Arnold, the department's research professor. Together, they built three different pilot plants to perfect the patented "Iowa State College process." The technology promised to revolutionize the soybean extraction industry because it did not use flammable hexane, a solvent that had caused plant explosions and deaths around the world. Instead, the Iowa State process used trichloroethylene, a chemical dating back to the 1920s in Europe and sold by DuPont.

When the time came to commercialize the idea, Iowa State and Crown Iron Works became partners, thanks to banker Lee Luick's introduction. "After all angles had been examined by the company," Clifford Anderson remembered, "licensing agreements were made with the Research Foundation of Iowa State College so that we could manufacture and sell these newly developed extraction plants."

Anderson announced Crown's exclusive agreement with Iowa State College and its own successful pilot plant testing in early November 1948. Crown went to work refining the new process and equipment involved. "We hired Mr. Luick and rehired Frank Scofield—a previous

EXTRACTOR

MEAL DRYING TUBES

VAPOR SEAL

MEAL COOLER

COUNTER FLOW EXTRACTION

PRIMARY MISCELLA TANK

SECONDARY MISCELLA TANK

FILTER

FILTER PUMP

Crown Iron Soy Bean Plant   *Minneapolis 1949*

tried-and-true estimator-engineer," Anderson said. Both of them served as salesmen for the new extraction process. Other key people—Frank Austin, Al Kaiser and Joe Givens—lent their engineering and operations expertise.

Austin, a college friend of Clifford Anderson, joined Crown in 1942, bringing his expertise with handling government contracts during World War II and later helping lead the company through conversion to the peacetime economy. A seasoned businessman and engineer, Austin became a company officer and director.

A native of rural Truman, Minnesota, Al Kaiser graduated from North High School at the start of the Great Depression, he managed grocery stores for National Tea Company and learned about drafting at Diamond Iron Company. Clifford Anderson recruited Kaiser away from Diamond Iron before World War II.

Joe Givens, a native of Northfield, Minnesota, with a degree from Carleton College, worked on the Manhattan Project in Oak Ridge, Tennessee, in the final steps leading to isolation of uranium 235 for the atom bomb. After the war, he was a management trainee at Cargill, before joining Crown Iron Works in 1948.

"I worked with Al Kaiser to build Crown's pilot plant and develop machinery for processing soybeans," Givens remembers. "The goal was to have machinery that processed 25 tons of soybeans daily and do it safely. Al was a natural, brainy guy. He wasn't formally educated in engineering, but he was talented. He was gifted mechanically and he was an excellent draftsman."

Crown bore the cost of experimentation and testing. "Our first plant between Quincy Street and Central Avenue in Northeast Minneapolis was quite successful, although it produced

(Upper Left): By 1949, Crown had refined the Iowa State College soybean extraction process and built its successful pilot plant (right) on Quincy Street in Northeast Minneapolis.

"Drs. Sweeney and Arnold early in their research realized the mechanical method of removing oil from soybeans . . . was inefficient and wasteful."

Minneapolis Star,
November 9, 1948

Hail the "Monarch of Manchuria"

> "The soybean came to this country as a stray immigrant about 1804. It took us over a hundred years to realize its value."

Mildred Lager, *The Useful Soybean*

No question about it. The Yanks were late in appreciating the value of what author Mildred Lager called "the little round bean" with a "spectacular and gigantic role" in worldwide agriculture. Long before Americans wised up to the worth of soybeans in the 1930s, the British and Germans were extracting oil from soybeans and other seeds two centuries earlier. Innovators in the Orient capitalized on the bean centuries before the Europeans to stave off starvation (historians called the soybean "the monarch of Manchuria").

"Presses of one sort or another had been in use for thousands of years to obtain oil from seeds, and these utilized various means for applying pressure to the material being processed," Warren Goss, an agricultural guru from Pillsbury Company, wrote in Soybean Digest in December 1950. "In some primitive types, for example, a system of levers was used and, in later models, winches and chains or ropes were incorporated. Others employed screws to apply pressure."

Debunking Devils

Historians credit Simon-Rosedowns of Hull, England, with the earliest invention of oilseed-crushing machinery in Europe, around 1777. The firm made a hydraulic press for a company in China around 1868 so advanced that locals considered it the work of devils and refused to venture into the plant. Meanwhile, Krupp Industrie-und Stahlbau in Harburg, Germany, began producing hydraulic machinery for oilseed processing around 1870.

At its Hull, England, site Simon-Rosedowns began producing batch solvent extraction equipment as early as 1898, but it was apparently the Germans—after World War I—who introduced the first continuous solvent extractors: "The end of World War I left Germany with a shortage of fats and oils as well as animal feedstuffs," The Journal of the American Oil Chemists' Society reported in 1977. "The Germans began to seek better ways to get the most out of their imported Manchurian soybeans." German inventors introduced two, continuous solvent extractors, the Bollman basket extractor in 1919 and the Hildebrant or U-tube extractor in 1934.

Simon-Rosedowns introduced its first continuous solvent extractor around 1949 and about the same time, DeSmet S.A. in Belgium and Lurgi Apparate-Technik GmbH of Germany started marketing their own solvent extraction processes.

The Yanks Catch Up

Meanwhile, in America, processing soybeans came in fits and starts.

Among the hydraulic presses, Pillsbury's Goss said the screw press, called the Expeller (trademark of Anderson International), became the most popular way to extract oil from seeds in the United States. Eventually, that popular method was eclipsed by solvent extraction, particularly with soybeans, though progress was slow.

"The pioneering efforts at solvent extraction in the United States may be likened to Leif Erikson's discovery of America, and the initial efforts did not produce lasting developments," said Goss. "Americans had little economic incentive to use solvent extraction before the 1930s."

Just the same, a solvent extraction plant for cottonseeds at Southport Mills, New Orleans, apparently ran on aviation gasoline and later benzene in 1917 and, as early as 1923, the Pratt County Soybean Cooperative operated a batch plant in Monticello, Illinois. Still another venture turned up in Norfolk, Virginia, in 1924, run by Eastern Cotton Oil Company. But none of these efforts proved profitable.

Long before America entered World War II, the most publicized soybean extraction efforts focused on Henry Ford's experiments and the work of two American companies—ADM and Glidden—both importing German equipment to produce a remarkable 100 tons of soybean meal and oil daily in 1934. Close behind them would be Central Soya, Honeymead Products and Procter & Gamble, who were supplied processing equipment by Blaw-Knox, French Oil Mill, Allis Chalmers and—before too long—an unlikely candidate, Crown Iron Works.

a rather sizeable loss, $50,000," Clifford Anderson said. Even so, he reasoned, the plant "proved the practicability of the process."

The Minneapolis Star newspaper ran the story about Crown's new venture on its front page on November 9, 1948: "Blast Danger Cut," the headline declared, "New Process Takes Soy Mill to Farm Area."

"The plant is designed to be set up in the heart of the soybean producing area, like the grist mill of pioneer days, and return oil meal feed to the farmer at big savings in freight costs," the newspaper said.

The *Industrial Supply Expeditor,* a trade journal, explained the advantages of Crown's plants further, "One of the great problems in soybean marketing has been the cost of transporting the beans from the farms to distant plants for processing the oil and meal. There have been two methods for extracting the oil, which comprises about 20 percent of the bean. One method is by mechanical-hydraulic pressing processes; the other, by chemical solvents. Both require large plants and, while the chemical process was more efficient, it was also dangerous, as the petroleum solvent used was highly explosive and flammable. Crown now builds a plant with 25 tons-per-day capacity which can be installed at any country elevator . . ."

Crown's Next Near-Death Experience

Crown had its new, big product. Never mind that it had nothing to do with ornamental iron or structural steel. It tapped Crown's knack for fabricating metals and assembling them into a functioning form. It called on Crown's growing expertise in engineering and its inherent talent for invention, experimentation and tweaking good ideas until they prove useful.

Crown sold its pilot plant and reinstalled it in Glencoe, Minnesota, for The Farmers & Merchants Milling Company in late 1949. Within two years, Crown also started plants in Blooming Prairie, Minnesota; Grand Forks, North Dakota; Fremont, Nebraska; and Millsboro, Delaware.

It was "all systems go"—until the rumors started in late 1950. "We received some very disquieting reports that cattle were apparently being poisoned and dying of a mysterious and unexplained malady," Clifford Anderson said. "It could only be traced to the meal produced by our plants. There was no proof to the contrary and the plants had to stop production until the problem could be solved."

There had been obscure warnings of potential danger. Cattle in England and France had died with a bloody nose disease linked to soybean meal extracted with trichloroethylene more than 30 years earlier. But other, similar operations had no such trouble. To make sure, Drs. Arnold and Sweeney had sent their soybean meal to Cornell University, where it was tested on two sets of cattle, doing no harm.

Nevertheless, Crown was in deep trouble. When farmers filed lawsuits against the mills using Crown equipment, those suits passed to Crown. "To make things worse," Anderson said, "we had contracted for other plants and all of our customers were holding up payments. Dozens of cows—and some sheep—died. All of these animals were 'prize winners,' 'best of the herd,' and very expensive."

"Soybean oil extraction plants of a new type, using a non-explosive and non-flammable solvent, and designed in relatively small sizes for serving localized areas of production, are to be placed on the machinery market soon by Crown Iron Works of Minneapolis, Minn."

**Chemurgic Digest,
December 1948**

"We were in the middle, holding the bag. Our plight was a desperate one. Bankruptcy was coming our way."

Clifford H. Anderson

The claims started climbing into the six figures, but Crown had only $25,000 in product liability insurance and no reserves to carry the company through this catastrophe. "Iowa State College had no explanation for the poisonous product," Anderson said, "and neither did E. I. DuPont." Crown would later discover that peculiar conditions with wet soybeans and occasional careless control of plant operations in a plant using trichloroethylene combined to produce the toxic meal.

In a journey reminiscent of the war years when Crown was on the ropes with a contract deadline and no raw materials, Clifford Anderson headed east to Wilmington, Delaware, home of DuPont. He made a personal appeal to the company's president, D. O. Notman. "I will always remember this episode for it illustrates what one strong man in a large company can do to assist a smaller company when it is in trouble," Anderson said. Against the strong objections of his legal department, Notman understood Crown's plight and agreed to assume half of the cattle death losses, as long as Iowa State helped, too. Anderson promised that Crown would settle all the lawsuits without involving DuPont. Iowa State anted up $50,000—all it had in its foundation at the time—and Notman ultimately sent Crown a check for more than $200,000.

Notman's sense of fairness toward a small business having little leverage with an industrial behemoth affected Clifford Anderson deeply. It confirmed his belief that integrity in life and business is the gold standard. His gratitude was great: "Without DuPont's help," Anderson said, "there is very little doubt that we would have been forced into bankruptcy."

After all the legal claims were settled, the losses totaled more than $250,000 and Crown paid still more to remove and rebuild the plants involved, using the time-tested, but flammable, hexane. The whole incident cost Crown time and its reputation for some years, but the company persevered in the extraction business—a decision it would never regret.

Even during the worst of times for Crown, Al Kaiser, the company's chief mechanical engineer, never lost heart. He played a key role in refining Crown's continuous solvent extraction system and he was an effective salesman, too. Kaiser persevered through Crown's cattle poisoning debacle and he worked hard to rebuild the company's presence in the solvent extraction industry.

Among many accomplishments, Kaiser is credited with designing an air-cooled oilseed processing plant for Cooperative Vegetable Oils Ltd. in Altona, Manitoba, Canada, where water was scarce and expensive. "Never having heard the phrase, 'It can't be done,' the engineers of Crown Iron Works did it," the local newspaper declared.

By the mid-1950s, Crown had re-grouped, re-tooled and re-engineered its solvent extraction process—at the same time the U.S. Department of Agriculture reported that American farmers had more acreage planted in soybeans than ever before in history and soybean exports were expected to reach 80 million bushels in a single year (1957).

Clifford Anderson kept his own copy of *The Useful Soybean* on his office shelf. It was inscribed by Lee Luick and dated 1947: "Dear Clifford, Please accept this book with my compliments . . . ," Luick's inscription read. "Now that you are definitely in the soybean business, I am sure the facts unfolded will be of interest and value to you. I also trust you will share with Mrs. Anderson as an assist for her important job of feeding your (always) hungry boys!"

> "Al Kaiser was part designer, part patent-holder, part salesman and start-up crew. He covered every facet of the extraction business."
>
> **Bill Kratochwill, veteran Crown employee and fellow inventor**

> "We are on the threshold of big things."
>
> **George Strayer, executive vice president, American Soybean Association, September 1957**

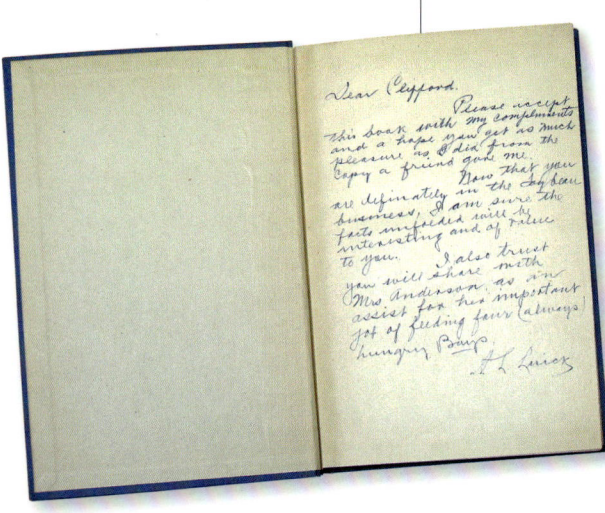

Nurturing Ideas

> "Crown Iron Works proved that ideas are as important as mechanical dexterity in the fulfillment of any assignment."
>
> *Minneapolis Sunday Tribune*, July 7, 1957

Frank Austin, Crown's vice president of engineering, knew how to nurture innovation. He looked the other way when some of his best and brightest broke a few rules. "Al Kaiser and I went out every morning around 10 a.m. for coffee," says Joe Givens, whose patented ideas helped Crown establish itself as a solvent extraction leader. "Nobody approved of it, but Frank let us do it. We were inventing machinery and we were having a good time doing it."

Givens went on to supervise start-up of Crown's first extraction plants and when he landed in Dawson, Minnesota, in 1951, he found his home and a plant that became his laboratory. Dawson's first soybean processing plant—championed by the local chamber of commerce—held huge promise for the little western Minnesota farming town. But the designated manager, an ex-farmer without technical education, was unprepared when rumors of dying cattle reached Dawson's city farmers.

Would the city's investment in this 25 tons-per-day plant go south? On Christmas Eve, the officers of Dawson Farmers Elevator Company called Givens with a plea—they asked him to become their full-time manager.

For Crown and Givens, it was a timely opportunity. Givens, an inventor and tinkerer at heart, had a challenge and Crown had an agreement to pay Givens for his process improvements. "I believed that by changing the method of desolventizing the trichloroethylene-extracted soybean meal, the toxicity problem could be solved," Givens remembers. "The changes were quite simple and they involved changing how the meal was toasted." Instead of cooking the meal with a pressure cooker, Givens used live steam. When Givens tested the meal on local calves, they thrived. "Dawson soybean meal never did have any toxicity claims against it," he says.

Even so, they had to sell their soy meal at a discount, because of the cattle deaths, until 1953, when Givens and Crown converted the Dawson Mills plant to use hexane solvent. Within one year, the demand for soybean oil and meal was so strong that Givens hired Crown to supply another extractor, increasing the daily processing tonnage from 25 to 80 tons.

Over the years, Givens lent his creativity to improving Crown's extractors. He also contributed to improving the company's desolventizing equipment and hot dehulling (a process used in preparation of the soybean). Givens' creativity and follow-through helped give Crown the competitive edge it needed to go head-to-head with older and more experienced European companies.

During Givens' tenure as manager of Dawson Mills, from 1952 to 1981, the plant boosted its daily processing capacity from 25 tons to 1,500 tons and multiplied its soybean products from simple meal and oil to about 20 different products. Sales increased from less than $1 million to about $100 million a year. The little town of Dawson enjoyed the fruits of that growth with more jobs, investment and economic activity.

One of Crown's earliest contracts was this 600 tons-per-day soybean extractor for Dawson Farmers Elevator Company in Dawson, Minnesota.

Teamwork for "Curious Dummies"

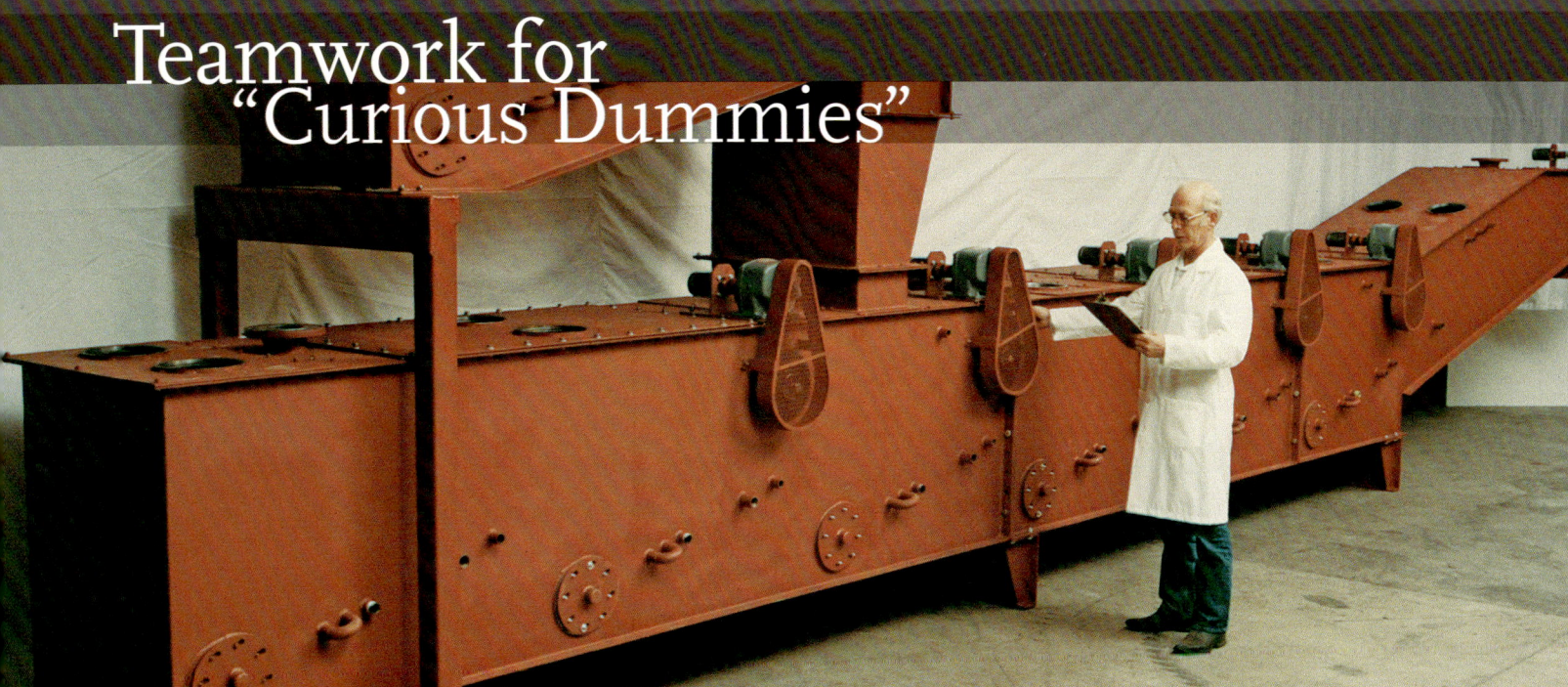

Bill Kratochwill was the classic inventor: a farm kid from southwestern Wisconsin who learned to be resourceful at an early age—solving problems and fixing machinery with ingenuity and baling wire. When he joined Crown Iron Works in 1965, Clifford Anderson assigned him to the manufacturing shop, where Kratochwill built Crown's first desolventizer-toaster (DT), a process that removes hexane from the oilseed flakes or cake, after the oil has been removed, and completes the toasting process. Building on the original DT invented by Central Soya about 15 years earlier, Kratochwill put his own practical spin on the process: "It seemed like a pretty good-sized piece of machinery for a little farm boy like me," Kratochwill, the holder of four Crown patents, recalls. The DT was bound for an oilseed processing plant in Manitoba, Canada.

For years afterward, Kratochwill lent his ingenuity to more than 40 plant start-ups with Crown equipment in countries around the world, including the Philippines and Guatemala.

He remembers the lean decade of the 1960s: "I did everything from sweeping the floors to scrubbing the pots to shaking hands with the president . . . and everything in between," he says. "Crown had a skeleton crew."

Kratochwill is co-inventor on two patents linked to Crown's first hot dehulling process, a key step in preparation of oilseeds. Hot dehulling is the fastest, most efficient process in preparing oilseeds for extraction and Crown was working hard to develop its own process. Kratochwill's other two patents are related to Crown's Model IV Extractor and the company's Down Draft Desolventizer.

Bill Kratochwill (above) holds the U.S. patent for this Crown Model IV Extractor.

> "When Crown had its own manufacturing operation, we had mechanical folks in the plant who liked to tinker. They were our innovators in the old days."
>
> **Clifford I. Anderson**

In the 1960s—as it is today—successful innovation boils down to teamwork, says Kratochwill. "It doesn't make any difference if you're making toothbrushes or a solvent plant," he says. "At Crown, we had a bunch of curious dummies. You see, you have to have a bunch of curious people to start with. And, they need to know what they can't do...as well as know what they can do. They communicate a lot back and forth. They figure out all kinds of different circumstances that excite their imaginations. Together, they solve their problems."

There's another key factor in this teamwork equation: ego. "Ego has to take a back seat," says Kratochwill, who believes Crown is a company surprisingly low on the ego quotient. "No one can feel he's smarter or better than the other guy."

Crown Iron Works Plans Open House on 75th Year

Crown Iron Works Co., founded in 1878 as a blacksmith shop, will hold open house Friday to celebrate its 75th anniversary.

Guests will see workmen fabricating ornamental metal work, structural steel and mechanical equipment and witness one of the most colorful operations of a foundry—pouring of iron and aluminum castings.

More Cause for Reinvention

In 1953, when Crown Iron Works celebrated 75 years in business, the company seemed strong. "Guests will see workmen fabricating ornamental metal work, structural steel and mechanical equipment and witness one of the most colorful operations of a foundry—pouring of iron and aluminum castings," the *Minneapolis Star* reported on May 13, describing the company's celebration open house. But behind the industrial sparks and publicity, Crown watched its long-standing businesses falter.

For decades, Crown had manufactured and sold a coal stoker for homes and businesses, but—by the 1950s—gas, a cleaner heat source, was supplanting coal. Crown discontinued its stoker business and turned to making repair parts and feed screws, a key component of the old stokers. The company invented a new, less expensive way to make a stronger feed screw using steel and, later, stainless steel. Before too long, Crown was offering feed (or conveyer) screws in sizes ranging from smaller than one inch up to several feet in diameter and 85 feet long. Crown's sales grew yearly through the 1950s and 1960s and machinery manufacturers all over the United States were their customers.

Ornamental iron fell from favor, too. "More and more, the ornamental iron work was changing from castings to extruded aluminum sections and rolled shapes being produced by Alcoa, Reynolds, Kawneer and the specialty steel section mills," Clifford Anderson said. "Architecturally, things were being streamlined to make them both cheaper and more functional, with less ornamentation. The problem of selling ornamental iron was becoming more and more difficult for us." Crown landed a few big contracts—stainless steel "ribs" supporting glass walls on the Prudential Insurance Company's North Central home office and American Hardware Mutual Insurance Company in the mid-1950s—but the bigger mills were eager to control the business and small players like Crown couldn't compete.

With ornamental castings for buildings no longer in demand architecturally, Crown's foundry shifted to making castings for other machinery manufacturers. Crown took orders from electrical machinery manufacturers, rock crushing companies and pump makers. The competition was fierce: "There were 27 companies in the Twin Cities, all fighting for this kind

For decades, Crown manufactured and successfully sold coal stokers for homes and businesses.

"The minds and metals to solve your problems."
Crown advertising slogan in the 1950s

(Upper Left): Crown provided structural steel for the Twin Cities' Metropolitan Stadium in 1957. Here, Clifford Anderson (center) is pictured with stadium planners. (Upper Right): Crown fabricated a sculpture for one of the nation's first covered shopping centers, Southdale, in Edina, Minnesota.

of business," Anderson said. "Castings were offered at cost or below. It was a continual battle to sell castings at low prices and then attempt to make a profit on them." There were two factors at work: the cost of labor and equipment. Crown invested $150,000 to upgrade its foundry, making it more efficient and competitive, but the combination of wages and overhead was crippling. Crown's actual costs were sometimes twice its project estimates and losses heaped upon losses with each contract.

No Ducking the Facts

Crown marched through the 1950s and 1960s with a few signature projects: structural steel for the Twin Cities' Metropolitan Stadium and the entire 1,650-foot outfield fence; a tall fabricated sculpture in the form of a birdcage at Southdale, one of the nation's first enclosed shopping centers; the first skyway, built over Marquette Avenue at the North Star Center; and the world's first all-steel band shell. But by 1961, 1962 and 1963, Crown was posting repeated losses totaling $852,000 in those three years.

Crown was on the ropes. Again.

"There was no point in ducking any basic facts," Clifford Anderson wrote. Working with Don Rittenhouse, the company's vice president and treasurer, Anderson took a candid look at his company. "Our company was imperiled with a real threat of collapse into bankruptcy," Anderson said. Crown needed nothing short of radical surgery.

"We closed our foundry, our largest department with 65 employees, in September 1963," Anderson recounted. "We then shut down the next largest department, ornamental iron, with over 30 workers. We shifted as many mechanics as possible into remaining jobs, governed by

seniority and ability. But after the cut-back, our total employment dropped to 72, from 211 the year before."

Crown sold its foundry equipment and machine tools to generate cash and retained four departments—structural steel, extraction, feed screws and fence sales. Clifford Anderson appealed to Twin City Federal Savings & Loan Association to increase its mortgage on Crown's Tyler Street plant from $100,000 to $250,000. The *Minneapolis Star* newspaper reported the effort: "Crown Prunes Way to Profit." Quoting Clifford Anderson on the new strategy: "We're re-emphasizing the profitable lines with the intention of making them the backbone of our activity."

Though this decade was dubbed the "Go-Go Sixties"—reflecting new business start-ups, glamour stocks and record trading on America's stock exchanges—Crown didn't join the party. With its tough decisions in the early 1960s, Crown managed to put together four profitable years from 1964 through 1967, but the decade would end with ever-increasing losses.

It was a sobering welcome for Crown's third generation of leaders, Clifford I. and George Anderson. Young Cliff, trained in business at the University of Minnesota, and George, a mechanical engineer from Stanford University, shared the same values, but their interests and talents were divergent.

When Cliff joined Crown in 1965 and George's summer jobs as draftsman became full-time work after graduation in 1969, the Andersons would defy common wisdom about siblings in business. Rather than butt heads and compete for control, their working partnership—based on respect, communication and clear boundaries—defined Crown's corporate culture for more than 40 years. 👑

(Upper Left): Crown's structural steel was part of the first "skyway" built in downtown Minneapolis crossing Marquette Avenue at the North Star Center and (Upper Right) the United Airlines Building at the Minneapolis-Saint Paul International Airport. Clifford H. Anderson is pictured.

"Crown was a family type of place where everybody took more than their share of responsibility. Those years were hard."

Marian Berglund, Crown employee, 1947–1987

Duck Boats and Minnie's Makeover

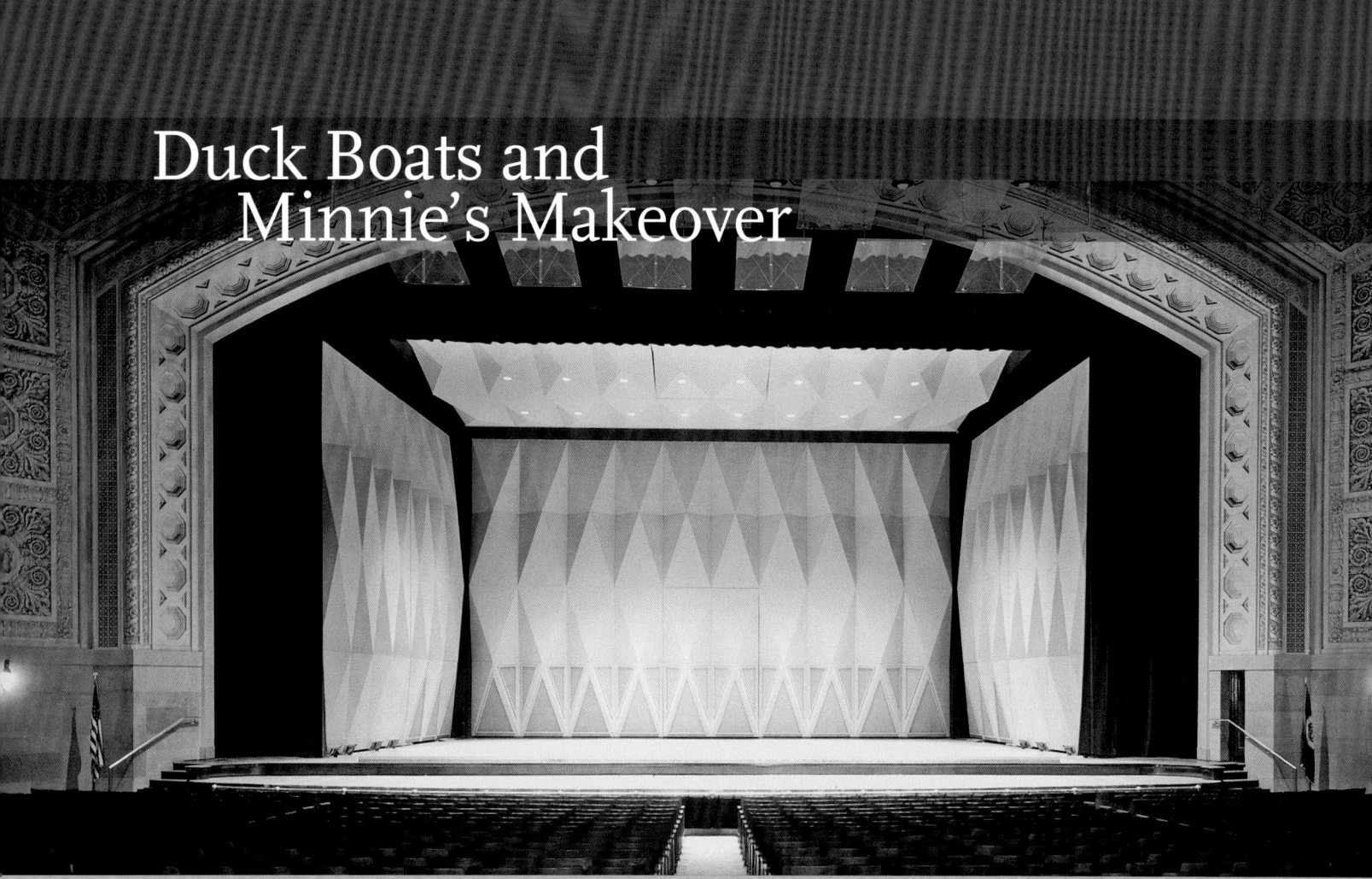

"We've been involved in unusual jobs before," Clifford Anderson told popular newspaper columnist George Grim of the Minneapolis Morning Tribune, in September 1961, "but this job makes the others almost commonplace."

Crown was building a 30-ton-plus, all-steel stage intended to make acoustical history for what was then called the Minneapolis Symphony Orchestra. "Walk through Crown's plant and you see parts of the new orchestral shell everywhere," reported Grim, a columnist who liked to report from on the scene. He described "15-foot-high sections that look like a metal fabricator's version of duck boats." These were welded and fitted into an enormous back wall designed to send the sound out to the Northrop Auditorium audience on the University of Minnesota campus.

Grim detailed the huge components of the revolutionary stage, all being fabricated by Crown: "Rolling side panels, a seven-ton steel ceiling that slowly hinges down via electric winch, 'wings' of transparent plastic that bridge over the top of the proscenium arch."

Crown worked at double time. "We're trying to accomplish six months' work in weeks," Crown's supervisor Earl Keller said, almost breathless.

Meanwhile, in another part of Crown's headquarters on Tyler Street in Northeast Minneapolis, Minerva had just emerged from her once-every-sixty-years makeover. "You remember Minnie," columnist Grim said. "In her flowing classic drapery, she stood in her niche above Hennepin Avenue at the public library for some 68 years. Her fingers kept turning the page of a book. . . . Her friends were the pigeons."

When the new Minneapolis Public Library opened, Minerva was to be moved to a pedestal, softly lit and indoors. "For months, she's been flat on her back, the covers pulled up," said Grim, "at Crown Iron. Three weeks ago, they stood her up—all 1,700 hollow bronze pounds of her—and began the cleaning up. Brawny workers, using brushes and mild soap, are going over and over her less-than-clean neck."

But there were limits to Minnie's makeover: the architects told Crown not to clean her back to a bright bronze. After all, Grim reported, they wanted her to look her age, with a patina of weathered green in the style of Miss Liberty.

Crown built an all-steel stage for Northrop Auditorium on the University of Minnesota campus that was hailed as an acoustical triumph.

Soybeans Supplant Steel

Eli Anderson's oldest grandson, Clifford Irving Anderson, had an engineering degree in his sights, just as his father did. He studied at the University of Minnesota Institute of Technology and landed in the Army Rocket and Guided Missile Agency in Huntsville, Alabama, for his obligatory military service. After he "mustered out" and returned to the University's business school, Anderson recognized his strengths more clearly: "While all things mechanical fascinated me," he told his fellow Minneapolis Rotarians in the early 1970s, "I much preferred to deal with the management part of business."

When Crown's third generation Anderson, Clifford I., became President in 1974, Crown was a $9.4 million company with oilseed processing generating 25 percent of annual sales. Here, Anderson (above, left) joins John Cross, executive director of P.O.S. Pilot Plant Corporation of Saskatoon, Saskatchewan, to inspect a pilot scale extraction plant. By the 1970s, Crown had already explored business outside the United States in Israel, Australia, Canada, Taiwan and even mainland China. When Crown reached its Centennial in 1978, another pivotal choice loomed large.

Anderson didn't go to work for Crown when he earned his business administration degree in 1962. "When I graduated, Crown had been in dire straits and my father was working mightily to save it," he says. "At that time, I had very little to offer as a manager. The last thing Crown needed at that stage was another salaried man, especially the inexperienced son of the boss!"

For decades, wise family business owners sent their offspring to work and prove themselves elsewhere before joining the family fold. Anderson took a job in production control at American Hoist & Derrick Company in Saint Paul, where he learned about large factory operations and managing multiple plants. When he joined Crown in 1965, he started in plant engineering and put Crown's Tyler Street facility into better operating shape after its recent lean years. Next, he moved into sales estimating, then materials and production control. Anderson's efforts boosted the company's volume and profits, giving his father confidence that his son was a deserving successor. The third-generation Anderson became Crown's President in 1974.

Four years earlier, in 1970, Anderson married Nancy Carlson, a native of Moline, Illinois, and a first-grade teacher at Kenwood School in Minneapolis. When the couple left on their honeymoon, the Crown factory was on strike. But when they returned, they heard the familiar noon factory whistle blow at Crown. To his relief, Cliff Anderson knew that the company he would soon lead was fully operating once again.

Anderson's honesty as a young business leader came through in early press accounts: "You have to stand back every once in a while to get the right perspective, but at the same time you can't bend back too far," he told Minnesota's *Corporate Report* magazine. "After all, people do understand who the company belongs to. You can't say, 'I'm just hired on here like everyone else'—that just isn't true, and there's no point in pretending it is."

Anderson would grow into his leadership role, becoming an executive who often kept his own counsel, communicated openly, prized integrity and rewarded initiative.

Another Mr. Wizard

George Anderson, Cliff's youngest brother by more than eight years, was a budding Mr. Wizard: excelling in math and speculating on the whereabouts of flying saucers with his equally brainy school chums. Even as a kid, George wanted to understand how things worked. "Cliff was a big influence on me," he says. "I'd watch him fixing his 1935 Mercedes or he'd take a distributor out of a worn-out Ford and say, 'It isn't built right' and I'd ask, 'What's wrong with it?' I started seeing how things were built, what their faults were and I began to think, I'd like to design things a little better someday." At age 14, Anderson is convinced, his father bought him an old car, just to introduce him to machinery.

Anderson graduated with Honors and Distinction in Physics and his teacher at Blake School, Harold Hodgkinson, urged him to attend Stanford University. Hodgkinson, a noted scholar himself, wrote a recommendation for his top student and George looked no further (his older brother Austin also attended Stanford and later became a Crown director).

While a student, Anderson worked summers at Crown as a beginning draftsman and pilot plant technician (these were promotions from lowly shop time-keeper and blueprint runner). Ulf Rinecker, a frugal, precise Crown engineer from Germany, became his mentor.

"Anderson is known as a shirt sleeves operator who spends as much time in the plant as he does in his president's office."

Corporate Report, May 1974

"As a little kid, George drew cut-away pictures of submarines. They looked a lot like his pictures of soybean extractors."

Clifford I. Anderson

Early in 1969, Anderson graduated with a degree in mechanical engineering and was president of the student chapter of the American Society of Mechanical Engineers. That same year, Crown's Chief Engineer, Al Kaiser, died suddenly. "Dad called me and said, 'When you're done with classes, we could use your help.' I finished up, didn't stay for graduation and came back to work full-time at Crown.

A few years later in 1973, he married Barbara Kredit, a high school Spanish teacher from Saint Paul.

Anderson would mature in his career and become one of his industry's preeminent engineers, admired by colleagues and competitors.

A Heavy Metal Medley

By 1971—the same year Apollo 14 and 15 sent Americans to the moon—Crown Iron Works started posting a profit after three dreary years of losses. Through the 1970s, Crown's sales grew from $4.2 million annually to $15 million by the decade's end, and the company attributed most of its financial gains in the first five years to its stalwart core business: structural steel fabricating. Crown accomplished this in an era of rampant inflation, costly gas shortages and domestic angst triggered by violent opposition to the Vietnam War and President Richard Nixon's resignation from the presidency in 1974.

Crown built 13 of the Twin Cities' unique skyway bridges by the mid-1970s. Crown also landed the contract to fabricate structural steel trusses—up to 24 tons each—that became the skeleton of a new, four-story building joining Hennepin County General Hospital and Metropolitan Medical Center in downtown Minneapolis. In this unusual design, the trusses supporting each floor housed all of the hospital's extensive mechanical and electrical equipment, creating more open space on each floor. For Crown, the project required 270 tons of steel.

In 1974, Crown won the largest structural steel contract in its 96-year history, involving fabrication of 3,600 tons for a new, $25 million coal-handling facility in Superior, Wisconsin. Designed to receive coal by rail from Montana and transfer it to Great Lakes ships, the terminal, located on a 225-acre tract along the St. Louis River, took two years to build.

"George Anderson is probably the biggest asset in the oilseed business Crown has ever had. He's brilliant . . . and incredibly creative."

Leroy Venne, former Crown engineer and Cargill plant operations supervisor

Clifford I. (left) and George Anderson began working together at Crown in 1969. Over four decades, they became strong and resourceful partners and leaders.

"Crown was struggling when I came back from Stanford. I remember saying to Dad, 'Well, we've still got the house' and he said, 'I mortgaged that recently, too.'"

George Anderson

(Above, Left to Right): Crown was involved in major structural steel projects during the 1970s including a huge, coal-handling facility near Lake Superior, a new Health Sciences building on the University of Minnesota campus in Minneapolis and construction of the IDS Crystal Court in downtown Minneapolis.

"A fabricator and a steel maker are not the same thing. We do not 'make steel': we buy steel shapes and plates and manufacture things out of them. A good analogy is a tailor who makes clothes out of cloth, but he doesn't weave the cloth."

Clifford I. Anderson, Crown president

There were other structural steel projects worth crowing about in the 1970s: the Twin City Federal Bank, Lutheran Brotherhood headquarters and IDS Crystal Court in Minneapolis; the U.S. Postal Service Bulk Mail Facility in Eagan (like its name, a building of considerable bulk); the West Bank Library at the University of Minnesota, a new addition to the University's Health Sciences Building, additions to Sister Kenny and Abbott-Northwestern Hospitals in Minneapolis; structural metal for the Science Museum of Minnesota in Saint Paul; Park National Bank building in suburban St. Louis Park; and several buildings at Wold-Chamberlain Field (later renamed Minneapolis-Saint Paul International Airport).

But though Crown's Chairman, Clifford H. Anderson, declared that "buildings are always going to be built and steel is the basic structural material that holds them up," Crown was in a business crowded with competition and strangled by shrinking margins. Fortunately, Crown had oilseed processing know-how to fall back on—an industry with growing international promise.

Telling It Like It Is

When young Cliff Anderson became president in 1974, Crown Iron Works expected its oilseed processing business to generate 25 percent of the company's total $9.4 million annual sales. *Corporate Report* magazine said the 36-year-old Anderson was "just in time to see Crown shake off the doldrums and begin a pleasing climb. . . . Anderson is guiding the firm in what well may be its most challenging—and profitable—period to date."

In his first presentation as president, the young Anderson assured Crown employees of his deep connection to the company: "Please believe me when I tell you," he said, "that this company means more to me than an investment or a plant. It is a group of people who are highly

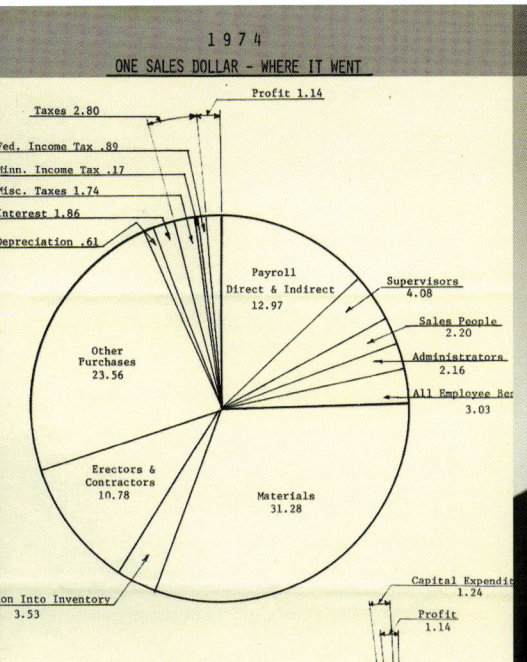

```
              1 9 7 4
      ONE SALES DOLLAR - WHERE IT WENT
                              Profit 1.14
    Taxes 2.80
Fed. Income Tax .89
Minn. Income Tax .17
Misc. Taxes 1.74
Interest 1.86
Depreciation .61
                                Supervisors
              Payroll           4.08
              Direct & Indirect
              12.97             Sales People
                                2.20
                                Administrators
  Other                         2.16
  Purchases                     All Employee Ber
  23.56                         3.03

     Erectors &
     Contractors      Materials
     10.78            31.28

                           Capital Expendi
ion Into Inventory         1.24
3.53                   Profit
                       1.14
```

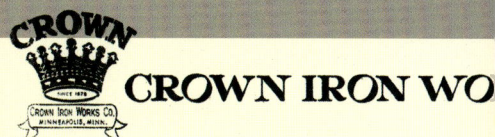
CROWN IRON WO

March 31, 1975

Dear Crown Employee and Family,

One of my objectives during my first year
to keep all of you informed about interest
company.

This year we will continue and improve on
Periodic letters such as this, in-plant me
will be continued.

In addition, on May 10, 1975, we will have
and their families. Some selected importa
be invited also.

regarded by me, just as they have been by my father and his father before him." To ease any anxiety, Anderson added with a humble grin, "My father has not retired. He is still Chairman of the Board and Chief Executive Officer and he will still be here in his office much of the time for advice . . . in case any of you were getting worried about this company's future!"

In his first years as president, Cliff Anderson focused on organization and stronger production controls. He moved Crown's custom-built screw conveyer business into space once occupied by Crown's defunct foundry and he trained his attention on manufacturing processes. Traditionally, Anderson says, "production control amounted to getting the drawings out into the shop and letting the foreman judge progress by eye." With Anderson's new system, each shop had standards to work against and a clear idea of how much time each task should take.

Anderson started writing a series of letters to employees and holding in-plant meetings, giving them candid updates on the company's progress and prospects. He told the truth and didn't sugar-coat the facts. He practiced "open book" management before there was a popular name for it, by producing pie charts showing employees where each sales dollar went, how much profit the company realized and how much it invested in people and capital improvements.

Anderson took an active role in working with Crown's union (United Electrical Radio and Machine Workers of America) alongside Sherry Robinson, who orchestrated union negotiations and follow-through. Anderson focused on the pension fund and learned the rudiments of union politics from a seasoned insider, Frank Rosen. He kept a file with a whimsical notation tucked inside: "An optimist is someone who labels a thin file folder, 'Union.'"

Anderson braced himself for the habitual fluctuations of his two biggest businesses. "Structural metals and soybean processing in particular are cyclical industries and Crown's

(Above): Young Cliff Anderson believed in "open book" management and regularly shared details of his company's performance with Crown employees and their families.

"Success, or even survival, in this business rests on our constant awareness of costs. It's the only way we can keep you working and the business solvent."

Clifford I. Anderson in a 1976 letter to Crown employees and their families

> "Crown was a little company in solvent extraction, but it became extraordinarily successful around the world, in virtually every country that had any role in agriculture."
>
> **Leroy Venne**

If it is possible to glamorize a Crown "cooker" (used in oilseed extraction), this photograph with Crown's good-natured sales secretary succeeded.

> "We never expected to grow as much as we did."
>
> **Glenn Brueske, Crown vice president of sales, 1960–1996**

heavy involvement in both means the company does not experience a steady progression of profits. There are good years and bad years," Cliff Anderson told *Corporate Report,* "and we hope they average out."

One of those bad years was 1975 and Anderson spoke with characteristic candor: "I was counting on the sales picture to brighten," he told employees, "because we had a lot of hot prospects going. Since then, nothing has materialized, although some of the prospects are still semi-hot. We still have a lot of structural work, but that can't make up for the current slump in conveyor screws and soybean processing equipment. I am sorry that my judgment was wrong about this. We're going to have to lay off about 10 percent of the plant and do some other belt-tightening as well." Anderson had avoided layoffs as long as he could, opting instead for shortened work weeks.

In the difficult times, Cliff Anderson's sincerity was apparent: "If you have suggestions, questions, gripes or just feel dissatisfied, see your supervisor. All of us in management are here to give you information. And—if you happen to get a chance—tell your supervisor about something good. Believe it or not, supervisors are human, too."

Selling the "Bigs"

Even in those hard years, Crown had reason to remain optimistic. Its extraction business showed real promise. "In every soybean mill going up in Taiwan, Australia, Israel or Canada, it's a good bet the machinery was manufactured in Minneapolis," the *Minneapolis Star* business section reported in the early 1970s. "The world-wide demand for protein has created new markets for the solvent extraction systems made by Crown Iron Works. And the need at home and abroad for sophisticated equipment—to process soybeans and other oilseed crops—has pushed company sales to between $5 million and $6 million in this industry annually."

By the early 1970s, Crown's solvent extraction business accounted for one-third of its sales and, by this time, Crown had sold one of the "bigs" in American agriculture: Cargill. Within the decade, they would also add blue chip Archer Daniels Midland (ADM) to their client roster.

The sale to Cargill wasn't easy. "I remember calling on the company with Clifford H. and the first thing someone said was, 'Oh, yeah, you're the guys who make those plants that kill cattle,'" Leroy Venne, a Crown engineer and later a Cargill plant operations engineer, remembers.

Young and Restless

Jeff Scott was a young man in his twenties when he left Anderson International of Cleveland, Ohio, packed up his family and drove west to Minnesota. "I think I was the first person Crown hired from out of state," says Scott, "and I wonder if folks were leery. I was coming from a city where the largest river was so polluted it caught on fire and our mayor was being impeached."

Scott's former employer had been Crown's export representative and occasional competitor in oilseed processing. Anderson International was considerably larger than Crown, but Scott says the company didn't have as much technological know-how: "Crown's solvent extraction technology was the standard for the future," says Scott. "The plants were getting bigger and more efficient."

Crown Iron Works became Scott's new home in 1977 and the company had tall aspirations, as Scott's sales meeting with his boss, Don Rittenhouse, demonstrated. "We sat down and reviewed a commission bonus program. In the first year, the sales projections started at $3.5 million dollars and went up to $12 million," says Scott. "But by year-end, we ended up at $2 million. I went to Don and said, 'Don, I'm a little confused. We're off this sales chart . . . on the wrong end.'"

"He looked at the financial goals and said, 'You know, Jeff, someday that's where we want to be.'"

Rather than catch a bus back to Cleveland, Jeff Scott embraced that sales challenge, winning his first $1 million sale to Nestlé in 1978, the day after Christmas. They wanted Crown to create an extractor that would remove oil from coffee, using that oil to introduce aroma into their freeze-dried brand.

Scott remembers meeting with Clifford H. Anderson, the company chairman. "He was a very proper and conservative guy," Scott recalls. "Whenever I got called into his office, my first thought was, 'Oh no, now what did I do?' But though he was strict, he was also very compassionate and kind." Scott got to know the senior Anderson best on the Crown fishing trips in northern Minnesota. "He was an avid outdoorsman, " Scott says, "and he loved to play poker. He seemed relaxed and happiest on those trips."

> "People at Crown are given a lot of freedom to succeed. The pay is good and our bonus system is based on total company profit. Every employee benefits because each of us contributes to the company's success."
>
> **Jeff Scott, vice president, sales and marketing**

Scott quickly moved beyond his first assignment by demonstrating his energy and creativity in forging strong working partnerships and promising joint ventures, building Crown's international presence and conceiving new business opportunities.

Looking back over more than 30 years with Crown, Scott is most proud of the company's growth and the opportunities he has had. "We went from perhaps four divisions with a primary strength in structural steel to becoming a process engineering company with nine divisions and global sales of $165 million. We went from building 1,500 tons-per-day plants to 9,000 tons-per-day. We are the only surviving U.S. supplier of our kind of equipment. We have offices all over the world. It's been a fantastic ride."

As it turned out, Anderson Sr. and Jim Spicola, head of Cargill's oilseed business, knew each other from Minneapolis social circles. In addition, Waldo Hardell, a member of the Crown Board of Directors, just happened to know a few influential people at Cargill.

"In 1971, Cliff asked to meet with Jim and he promised him it would be worth his time," Venne remembers. "I wasn't a salesman, but Cliff asked me to join him and back him up with technical information." Anderson appealed to what the companies had in common: Cargill and Crown were both local, family-owned companies, founded about the same time. Crown even supplied the larger and more successful Cargill with hull sections for shipbuilding during World War II. "Cliff told Jim, 'Given our shared histories, you owe it to us to at least look at what we have to offer,'" Venne says. "'Our expertise will make Cargill even better.' That was a nervy move. I remember thinking, 'My God, it's all over now.'"

Anderson and Venne described the design of Crown's extractor: far simpler and more efficient than those offered by Crown's many competitors. "Crown's process wasn't cheap," says Venne, "but it offered an economic advantage. It simply made a whole lot of sense."

The meeting was short and Spicola (who later became Cargill's president) was gracious. He saw Anderson and Venne to the door, saying, "We'll probably take a look at your product; it's interesting." Within a week, two Cargill engineers called Crown to see a demonstration plant and, within six months, Cargill purchased its first Crown extractor for a soybean processing plant in Cedar Rapids, Iowa. "Cargill wanted to run 1,500 tons in 24 hours and Crown's biggest extractor had a 700-ton capacity," says Venne. "Cargill put its old extractor and Crown's new one side by side. That sale was the opening for Crown. Other companies started to say, 'If Cargill's buying from Crown, they must be good.'"

Two years after that, in 1973, Crown posted its first $1 million sale in the solvent extraction business when Farmers Grain Dealers Association, a large Iowa cooperative, asked Crown to build its new 2,000 tons-per-day extraction plant in Sergeant Bluff, Iowa. "We had remodeled their old plant in Mason City and increased its capacity," Glenn Brueske, vice president of sales from 1960 to 1996, remembers.

"Can We Do It?"

In 1979, six years after Crown landed its Farmers Grain contract, an opportunity came from ADM. "We hadn't called on them yet," Brueske says. "Harry Ground was head of their soybean operation. He called and invited us to Decatur, the company's headquarters, the very next day."

Brueske and George Anderson jumped at the chance. Brueske remembers the start of the meeting with Ground and ADM's president, Jim Randall: "I remember Jim saying, 'Tell us about your equipment, but forget about that shallow bed and turning-the-bed-over pitch. We already know about it.' George and I laughed. He stole our best selling point."

Though Crown had less experience than some of its competitors, it had a simple and efficient extraction process adapted from the original Iowa State College design acquired nearly 30 years earlier. "We called it shallow bed," says Brueske, "and it was about two feet deep. Our competitors used cylinders about ten feet deep that rotated. In our system, the beans moved

---

This illustration was created in 1967 for Quincy Soybean Company, who later became a Crown customer. The whimsical approach illustrated a complex process in the early years of this industry when Crown was just beginning to build a customer base. The Rotocel wash tub represented a competitor's trademarked process, but Crown invented an improved way to handle soybeans.

"For our bigger competitors, oil seed processing was a cyclical business. When it went down, they had other interests to focus on. For us, it was big, no matter what. I think we just took it more seriously."

Cliff Anderson

"It was a new kind of machine. It was a product we'd never done before, a company we'd never sold to before. Our sales and engineering guys had reason to wonder, 'Can we do it?'"

George Anderson, Crown vice president, engineering

A Day in the Life of "Sam Soybean"

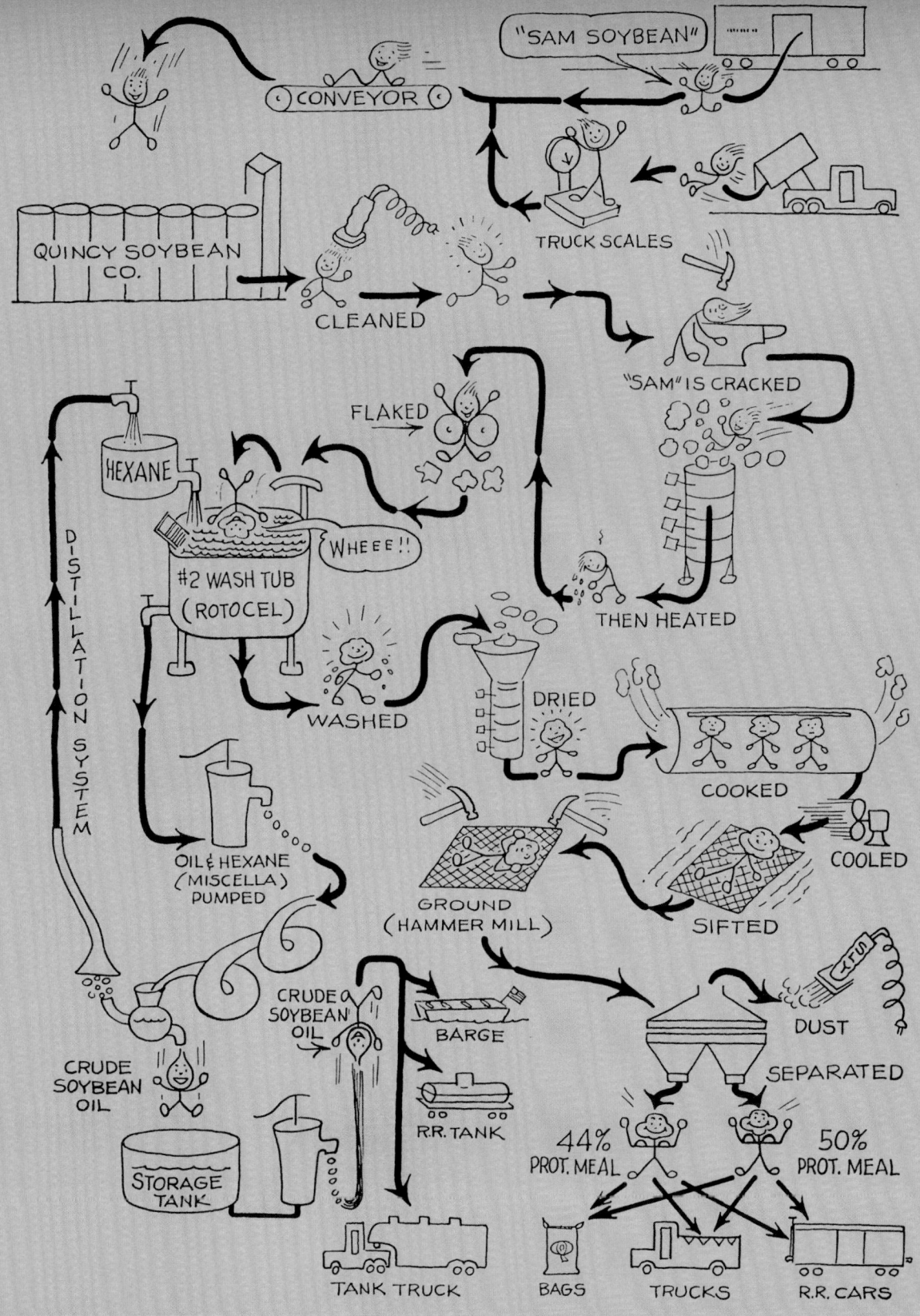

(Above, Left to Right): A Crown employee inspects an oilseed extractor sprocket hub, and another Crown employee uses this autoweld machine designed to secure a conveyor screw flighting, or helix, to the center pipe.

> "ADM's president said, 'Your machine isn't as big as your competitors'. They have more cubic feet than you do.' I remember answering, 'Our cubic feet are bigger than their cubic feet.' He glanced at me and very slightly smiled."
>
> **George Anderson**

along a loop in the shallow bed. It allowed for better solvent contact with the beans and better drainage. The system was also easier to maintain."

"ADM wanted an extractor and a desolventizer-toaster-dryer-cooler (DTDC) to process 1,200 tons-per-day of corn germ," George Anderson remembers. "The extractor was twice as big as anything we had built, with lots of stainless steel, for an oilseed we had never worked with. It was a major, new design that I had only made sketches of, but they didn't like our competitors' options.

"They wanted to order something we'd never built before to handle something we'd never run in our equipment," says Brueske. "They put a lot of confidence in us."

Within a week, Crown got its first order from ADM. "I think they liked our different approach," George remembers. "They knew that no one had a good machine for this work, so they might as well try us." ADM was willing to work with Crown on refinement of the system. "They helped us," says Anderson, "and that was the start of a great cooperative partnership that we've had with ADM ever since."

George Anderson used off-hours in his basement to convince himself that Crown could do the job they had just sold. "The machine was to be stainless and about 85 feet long, costing about $2 million. I didn't want us to screw it up," he recalls. Anderson built a model to test how well the mechanical chain would work, moving the corn germ around the much larger and redesigned machine. "I was home with the flu," he recalls, "and I built a three-foot-long extractor, making 60 little links out of scraps of aluminum and using electrical wire for pins. It was like being a kid again. A few people worried that the chain wouldn't work in this new design and I wanted to prove that it would. It didn't matter that it was a tiny scale model." Anderson brought his model into the office for a successful demonstration. "Then we built the real thing for ADM," he says.

Strategic Collaboration

Crown's association with Cargill, and later ADM, carried them overseas as these huge companies expanded their global business. By the 1970s, projects in Israel, Taiwan, Canada, Australia, the Philippines and Venezuela began to appear on Crown's project map.

Outside of plants in Canada dating back to the 1950s, Crown had its first foreign contract in Ashdod, Israel, in 1970. The relationship started with Daniel Chajuss, the son of Elias Chajuss, who invented Soy Protein Concentrate (SPC) in 1963. Chajuss' goal was to provide human and livestock food with high concentrations of digestable proteins.

Daniel Chajuss traveled to the Midwest to learn about agricultural processing methods and came upon Crown's extractor. Specifications in hand, Daniel returned to Israel and recommended the purchase to his father. This initial project led to decades of collaboration between Daniel Chajuss and Crown around the world. By 2007, more than 80 percent of the SPC plants in the world ran with Crown systems.

Unlike larger American companies that established their own international offices, Crown didn't have the staff or resources to support this kind of growth. Instead, they relied on relationships with two established companies with international ties: Anderson International, based in Cleveland, Ohio, and the Rosedowns division of Simon-Rosedowns in Hull, England. Both companies wanted to expand their oilseed extraction line and Crown's products were a good fit.

This collaboration was a smart strategy that would lead to more international relationships spanning many decades and continents. "Our market moves around the world—depending upon agricultural trends," says Cliff Anderson. "Crown needed an international presence and both Anderson and Rosedowns had strong international networks."

(Above, Left to Right): This Archimedes Screw Pump, produced by Crown, is capable of pumping 6.5 million gallons of water per hour. It was used in Texas City, Texas, after a hurricane pummeled the town. A Crown employee drills holes for stay bolts in a Crown DT (desolventizer-toaster).

"Our main export method has been to serve our global customers wherever they go and then expand our marketing in those new locations."
Cliff Anderson

Aging Aircraft and Blind Transport

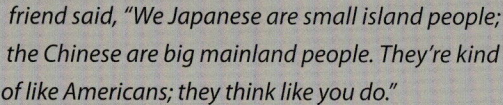

George Anderson was just 25 in 1972 when he packed his first passport, a couple of engineering books and spent nearly three months in Taiwan where almost no one he worked with spoke English. "The experience primed me for travel to mainland China a few years later," he says.

Anderson was to assist the Taiwanese in building and starting up a new soybean processing plant, and communication involved improvisation. To convey the limits of operating certain gauges, Anderson acted out the meaning with hand signals and put arrows on the thermometers with bright red nail polish.

In 1979, Anderson traveled to mainland China, this time with Glenn Brueske. (Communist leader Mao Tse-Tung's suffocating, 30-year grip on China had ended with his death in 1976.) A Japanese friend assured Anderson that they would like the mainland Chinese. When Anderson asked why the Chinese would be so likeable, his friend said, "We Japanese are small island people; the Chinese are big mainland people. They're kind of like Americans; they think like you do."

International travel in the 1970s to the Pacific Rim was rustic. Anderson remembers flying Northwest Airlines to Japan and then boarding an aging Chinese 707 with oil dripping from the engines and grass mats in the aisles for the final leg to Beijing, China. "We were served a small bird chopped up with the bone still in it and tea in a beat-up tin pot," he says. "The guy next to me, from General Motors, groused about being assigned the trip." (Anderson's trip home was aboard a fully modern Chinese 707 operated by the same airline.)

They arrived after midnight at a modest airport dominated by a huge picture of Mao and, beneath it, a sleeping customs official. Their Chinese hosts drove Anderson and Brueske to the old Imperial Hotel without headlights: "They couldn't afford the bulbs," Anderson says. "We dodged a man on a bicycle."

Things have improved a great deal in Taiwan and even more in mainland China, George Anderson says today, but in the 1970s, the contrasts were far greater. Looking back at his trip notes, he found these observations: "Mainland is the unbelievable 'Communist gray,' of nations where there is no joy in maintenance because private ownership of housing or business is illegal."

"Taiwan is a flash of colors and signs. Mainland is economically inefficient; Taiwan is the number one obvious success story. Mainland

9

> "Our baggage was never opened or searched. We were never hassled by any authority."

George Anderson in his 1979 trip report

(Above, Left to Right): George Anderson meets his first Taiwanese customer, the chairman of a local soybean processing plant, joined by his family. George and his father, Clifford H., tour the plant with Mr. Tung (left), an interpreter, and Mr. Han, chairman. George takes time out in 1979 to visit the Great Wall of China with local plant employees. Glenn Brueske (left) of Crown was an adventurous traveler with George Anderson in China. Anderson and Brueske visit a soybean processing plant in Mein Yang, near Wuhan, China. (Below Left, Right): George Anderson has a file of used passports. A 2002 advertisement extolls Crown's expertise in Chinese.

is feet, bikes, buses—except for the taxis and spotless black, chauffeured limousines of leadership; Taiwan is motorcycles, cars and airplanes."

Anderson and Brueske spent five days meeting with Chinese engineers and business people explaining—through interpreters—the American soybean processing industry, from preparation through extraction and refining. China needed to feed its large and growing population (plus its livestock) and America's technology offered a ready solution.

The pair from Crown brought a functioning model of an extractor with them, housed in a glass case. "At a break in the meeting, Glenn and I left to go back to our rooms," Anderson remembers. "On the way, we remembered a briefcase we left behind in the hotel conference room. We went back and found the Chinese delegation crouched over the model, the glass case off, measuring every segment like crazy. They were talking a mile a minute. When they saw us suddenly walk in the door, they pocketed their tape measures and just started walking around the room in dead silence. Boy, were they embarrassed!"

Crown would later learn that a dozen plants operating in China used knockoff versions of Crown equipment. It would take almost a decade before Crown landed its first project there, but savvy business people around the globe understood the long cultivation process steeped in ubiquitous Chinese green tea. "We didn't expect to write any orders from that first trip," Cliff Anderson told Corporate Report magazine in 1979, "but we've got the equipment, they need it, and they're interested."

For companies willing to invest their time and energy early in China, there would be commensurate returns, as Crown discovered in the 1980s, 1990s and 2000s.

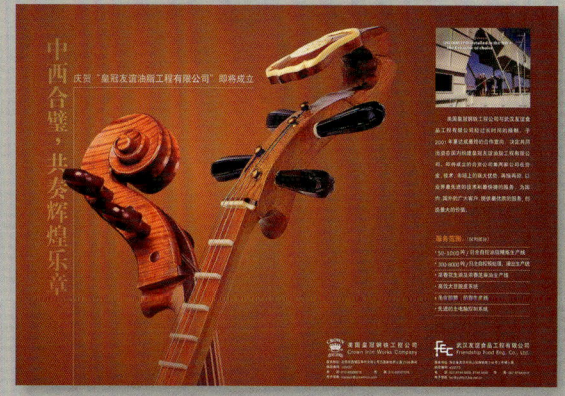

A Pivotal Choice

Though a centennial celebration is usually cause for celebration, Crown's 100th year in 1978 was mixed. Yes, few American companies actually survive 100 years in business and Crown joined this venerable minority. But Crown's balance sheet was doddering. Cliff Anderson told employees the distressing news: their company was operating in the red.

America's economy fell into another slump during the late 1980s with double-digit inflation and soaring prices at the gas pump. The construction industry had not rebounded as predicted and Crown had fewer large structural steel jobs. Though it had a backlog of medium-sized orders, the big ones were harder to land. Anderson described one order that characterized the market. "This order includes structural steel, plates and some conveyor sections and totals 509 tons," he said. "But unlike most jobs, we didn't bid on this one. Rather, we were offered the job at their price and told if they couldn't get it at the price they quoted, they would simply skip us. That's how tight this work is getting." From that point on, a resolute Anderson said Crown would reject all "no-profit" jobs.

Cliff Anderson, like his father before him, had to make a wrenching decision in 1981. Structural steel, the company's core business for decades, was no longer an option for Crown. Too much competition, narrow to no profit margins, the higher cost of doing business in Minnesota and a lagging economy were just a few factors in Anderson's analysis. Ironically, Crown had won a $1.8 million contract in 1980 to provide all the structural steel for the Metropolitan Sports Facilities' new Hubert H. Humphrey Metrodome, in downtown Minneapolis. But this high-visibility opportunity was no indication of market strength.

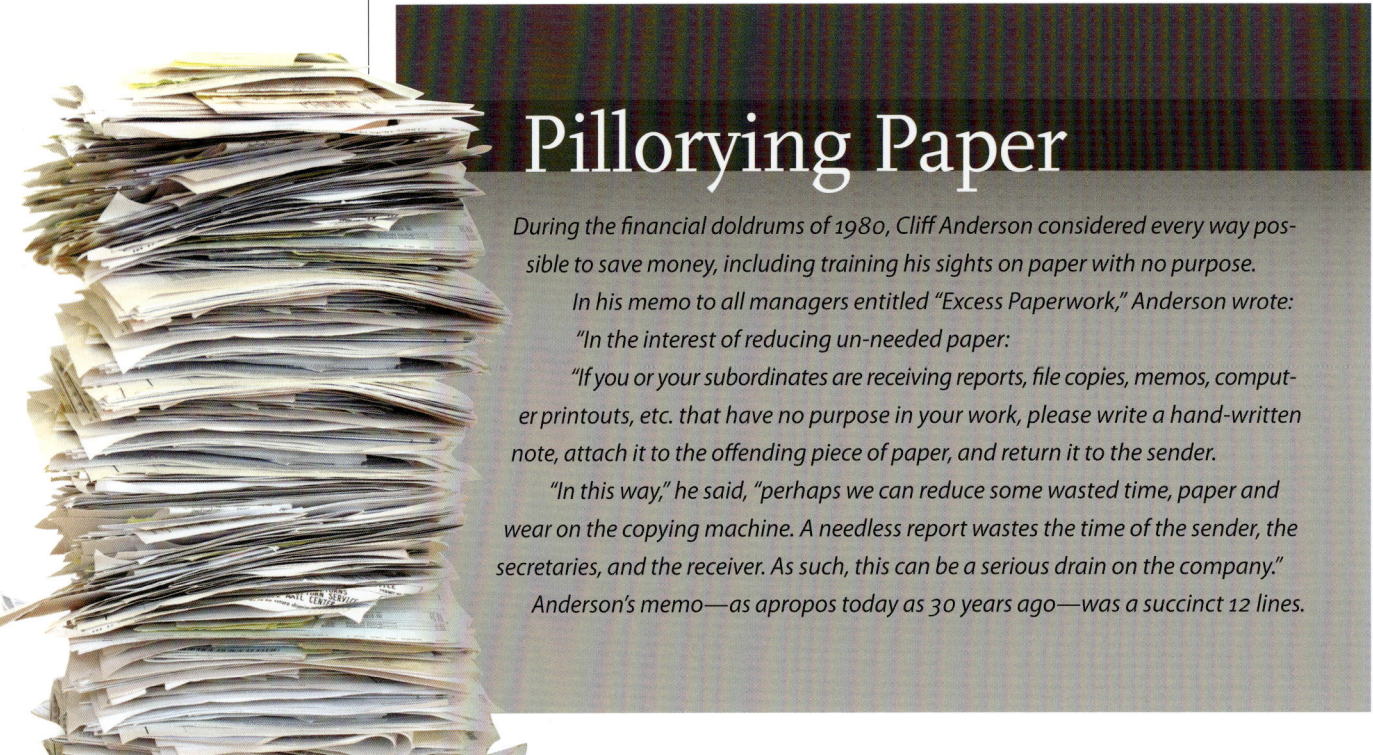

Pillorying Paper

During the financial doldrums of 1980, Cliff Anderson considered every way possible to save money, including training his sights on paper with no purpose.

In his memo to all managers entitled "Excess Paperwork," Anderson wrote:

"In the interest of reducing un-needed paper:

"If you or your subordinates are receiving reports, file copies, memos, computer printouts, etc. that have no purpose in your work, please write a hand-written note, attach it to the offending piece of paper, and return it to the sender.

"In this way," he said, "perhaps we can reduce some wasted time, paper and wear on the copying machine. A needless report wastes the time of the sender, the secretaries, and the receiver. As such, this can be a serious drain on the company."

Anderson's memo—as apropos today as 30 years ago—was a succinct 12 lines.

German Engineering, Yankee Ingenuity

DISCUSSIONS WITH SCHUMACHER
DATE: 3-28,29,30-1977
BY: GEORGE ANDERSON

1. ANCONA OPERATION: 3.0m (76.1 FT²) 1-1-1. 10½% FLAKES, VERY LOW HEXANE CARRYOVER, 11¾% MEAL, 250°F AIR, 385 SHORT TPD. FLAKES 0.012, WITH HULLS. EASY OPERATION AT 5.06 TPD/FT². DIFFICULTY MIGHT BE EXPECTED (320°,12½%) AT 6.50 TPD/FT².

2. SAMOR OPERATION: 3.5 m (103.6 FT²) 1-1-1. 10% FLAKES, VERY HIGH SOLVENT CARRYOVER, 12½-13% MEAL, 340°F AIR, 495-605 SHORT TPD. FLAKES 0.012, WITH HULLS. EASY OPERATION AT 4.78 TPD/FT².

Heinz Schumacher had a long and complex relationship with the United States. Schumacher was a young German glider pilot when he was shot down in North Africa during World War II. He was sent to a Louisiana POW camp, picking cotton until the war ended.

The next time he came to America was the mid-1970s when N. Hunt Moore, a veteran industry consultant, introduced Schumacher to Crown Iron Works. Ultimately, they penned a lucrative licensing agreement giving Crown the rights to sell Schumacher's superior DTDC for oilseed processing. In fact, Crown became Schumacher's first and only licensee in the United States.

"Heinz had sent a letter out to Crown and all of the big competitors in our industry offering licensing agreements," Cliff Anderson recalls. One rejected the offer because the DTDC needed refinement. Another chose to develop their own. Crown decided to work with Schumacher. "Like all German engineers, Heinz built his DTDC like a watch," says Anderson. "It was complex. Crown was able to simplify the process and take some cost out by working with Heinz. Ultimately, we outsold our competitors eight to one and Heinz became a fast friend."

When others gave Schumacher the cold shoulder, Anderson says, Crown was receptive and fair. That posture has been vital to the company's growth: "My advice to anyone who wants to be successful in product differentiation is to be generous to inventors," Anderson told a business audience in 1990. "Your reputation will bring others to your door."

Working with Schumacher was an unforgettable experience. George Anderson "shadowed" Schumacher on an early trip to Italy in 1976. "Heinz was a big, happy, volatile German who loved travel," Anderson remembers. "He was unconventional—an eccentric inventor type. He drove 110 miles an hour all the time, flat out in a big BMW. I remember him taking representatives from Cargill on a wild ride through Rome. They tried to follow him in a little Fiat 128."

Schumacher was a mentor to George Anderson and soon became a close friend. They worked together on refinements of the Schumacher DTDC and its second-generation version. "If we disagreed," Anderson recalls, "we'd argue just for fun. We never had a

time when one of us said or thought, 'I don't trust you.' We discussed every idea he had." Anderson's correspondence with Schumacher spans years and multiple file drawers. Letters detail measurements, tests and process refinements. "It's pretty obvious from looking at these letters that invention is mostly about sticking with something and following it through," says Anderson.

Not long before he died, Schumacher sent Anderson a set of new drawings for an invention that improved a key step in oilseed processing. He liked to keep Anderson on his toes: "He didn't want to tell me what it was for about a year. He liked to keep me guessing," says Anderson. "He said, 'This is so secret I won't even tell you.'"

Heinz Schumacher was an eccentric German inventor whose ideas gave Crown a clear advantage in world competition.

"Our mentoring phase became a friendship. Innovation is essentially play and Heinz and I loved to 'play' together: dreaming up new gadgets that fit the laws of nature and accomplished a task."

George Anderson

Schumacher nicknamed that invention "Super Clue" but eventually he shared the details with his young friend as they traveled to Holland to investigate how to improve the first commercial unit.

Even today, Anderson refers back to ideas he and Schumacher exchanged in their many letters and drawings spanning more than a decade.

PHONE DIALOGUE

1 (Name) GEA
Talked with Heinz (HAMBURG) of
☐ I called ☒ Party called   Time 3:00   Date 10-7-77
Subject DTDC
What said — Suggests 4 M & for 900 on DTDC or 157".
10% BIGGER. "SCARED OF DE HULLED". 150" may be a bit tight? Not sure, as he has not hit any limits. We compared various sizes, but not any limits.
What [party] said — on dehulled it is still 1.04 × D.T. volume with 6.5' bed runs for 900 tons only will run at 750-900 mm. He will write.

Action Needed

Anderson knew that. "We had to look hard at reality and ask ourselves, is this a flash, is it going to last?" Cliff Anderson asked. "We had seen the business come and go before.

"Getting out of structural steel meant we were signing the death warrant for our manufacturing business. We had two decisions to make: one, to stay in or get out of the structural steel business and, two, were we going to continue to be manufacturers?"

These decisions shook the basic underpinnings of Crown and Anderson knew it. Crown's Board of Directors insisted on careful analysis and he delivered. "They wanted me to show the facts of why I was pushing this decision," Anderson says. "I made up a thick loose-leaf notebook of facts and statistics. It was painful to review." Based on Anderson's presentation, Crown's experienced outside directors, Thomas Colwell, CEO of Colwell Press, and Martin Kellogg, CFO of Tenant Company, advocated action.

Clifford H. Anderson was, at best, ambivalent about the decision. "Dad wasn't averse to making big decisions," Anderson says. "He made them himself when Crown was almost broke and he closed the foundry." But this decision signaled a decisive end to a business the Anderson family knew well—one that still generated nearly 50 percent of Crown's annual sales (though increasingly unprofitable sales).

About 12 people in structural steel sales, engineering and drafting lost their jobs when the Metrodome job was completed and another 90 people left Crown a few years later when the union forced the company's hand and Crown shut down its shop, moving its screw conveyor operations to Cokato.

By 1982, Crown Iron Works was beginning its rebound. "Sales are poor," Anderson told Crown employees in March 1982, "but we have written just as much total volume as we had last year at this time. In fact, our conveyor screw department is 41 percent ahead and processing has booked 35 percent more. That is enough to replace the work sold by our structural division that we shut down last May." In fact, Crown Iron Works was now the equal of its arch competitor in the oilseed processing business, a company called French Oil Mill Machinery Company, of Ohio. And, in 1980, Crown sold its largest soybean processing contract to date: a 2,500 tons-per-day plant for Boone Valley Cooperative in Eagle Grove, Iowa.

Anderson reminded Crown's much-smaller workforce of the pluses that would carry the company forward: "Our quality is good and improving," he said. "We are delivering our products on time, as promised. We have a long history that our customers can rely on. Our banking relations are excellent, which enables us to finance the labor and material costs of large orders. You, our labor force, are certainly more ambitious than others. You have a good standard of living, you want to continue it and you have seen the connection between high productivity and the good life.

"Longer term," Anderson concluded, "I am optimistic." 👑

"It may surprise you, but price isn't everything. If it were, there wouldn't be a manufacturer left in this state. We have excellent products in our Processing Division backed by solid engineering and customer service."

Cliff Anderson in his Report to Employees, March 1982

Crown Timeline
1878–2008

1870s The "Swedish surge" swamps the Twin Cities. Record numbers leave their homeland and re-settle in Minneapolis and Saint Paul.

1878 August Malmsten and Andrew Nelson form Malmsten, Nelson and Company, manufacturers of machinery, bolts and tools, on the banks of the Mississippi River.

1882 Fire annihilates the little company. The owners scrape up enough cash to buy a little stone building at 113 Second Avenue Southeast.

1884 Malmsten and Nelson incorporate their venture as Crown Iron Works Company on January 6.

1888 Crown creates the cast and wrought iron for the Guaranty Loan Building, Minneapolis' first skyscraper at 12 stories.

1889 Dassel, Minnesota, farm boy Elias Anderson joins Crown as a $30-a-month bookkeeper. He ultimately becomes president and general manager.

1892 Presidential candidate Benjamin Harrison is selected at the huge Exposition Center, one block from Crown. He beats William McKinley for the nomination. (In 2008, the Republican Convention returns to the Twin Cities.)

ESTABLISHED 1878
CROWN IRON WORKS COMPANY

TYLER AND 13TH AVE. N.E.
MINNEAPOLIS, MINN.

IN REPLYING REFER TO

1903 Henry Ford founds Ford Motor Company.

1905 Fire again destroys a portion of Crown's plant. The company buys property on the outskirts of Minneapolis, at Thirteenth and Tyler Streets Northeast.

1906 The San Francisco earthquake kills 700 and causes $400 million in property loss.

1912 Woodrow Wilson wins the U.S. presidential election; Titanic sinks on her first voyage.

Elias Anderson

1913 The federal income tax is introduced in the United States.

1916 Crown wins its first structural steel contract for the St. James Hotel in Minneapolis.

1917 When America's first armed division arrives in France to join World War I, Crown joins the war effort by producing riveted steel for cargo vessels and inventing a better way to make barbed-wire "entanglement" fences to protect soldiers in trenches.

1923 U.S. President Warren Harding dies in office; Vice President Calvin Coolidge succeeds him.

1925 Madison Square Garden opens in New York City.

1926 Elias L. Anderson becomes president of Crown Iron Works.

1928 Crown celebrates 50 years in business with annual revenues reaching a record $1.5 million. Clifford H. Anderson, Elias' son, graduates from Stanford University and joins Crown's engineering department three years later.

Clifford Anderson on his way to college, 1926

Crown employees in 1937

1928 Herbert Hoover is elected America's 31st president.

1929 After hitting an all-time high of 381, the Dow plummets. Trading volume spikes at nearly 13 million shares and companies on the Big Board lose $30 billion in market value. The ten-year catastrophe called the Great Depression begins.

Soldiers erecting Crown screw posts.

1930s Crown posts yearly losses, cuts salaries and takes all paying work, no matter how little profit it generates.

1931 The *Star Spangled Banner* is officially designated as the National Anthem by act of Congress.

1932 Franklin D. Roosevelt wins the U.S. presidential election in a landslide.

1933 The 21st Amendment to the U.S. Constitution repeals prohibition.

1935 The AFL Union organizes Crown's workers.

1936 Margaret Mitchell wins the Pulitzer Prize for her novel *Gone With the Wind*.

1941 When Japan bombs Pearl Harbor on December 7, America joins World War II and Crown produces barbed-wire entanglement screw posts, metal decking and fittings for "pontoon bridges" and structural steel for ocean-going tankers and Army towboats.

1944 Crown wins its first of three Army-Navy "E" (for excellence) Production Awards.

PRODUCTION AWARD

CROWN IRON WORKS COMPANY

1945

1945 World War II ends in Europe. The United States drops atomic bombs on Hiroshima and Nagasaki; Japan surrenders.

1946 With the war's end, Crown sees all of its government business disappear and has no domestic customers waiting in the wings.

1946 Elias Anderson dies at age 76; his son, Clifford, age 42, succeeds him.

Clifford H. Anderson

1947 Crown Iron Works explores a partnership with Iowa State College to commercialize soybean oil extraction.

1947 Jackie Robinson becomes the first black to sign a Major League baseball contract.

One of Crown's many conveyer screws

1948 Crown builds its own pilot plant for testing Iowa State's technology and sells its first extraction plant to The Farmers & Merchants Milling Company of Glencoe, Minnesota, one year later.

1948 Harry S Truman is re-elected U.S. president.

1950s Ornamental iron falls from favor among architects and builders. Crown pursues its conveyer screw business, producing screws smaller than one-inch in diameter and others up to 85 feet long.

1950 Disquieting news of cattle deaths are linked to Crown's extracted soybean meal. Insurance claims climb into the six figures and Clifford H. Anderson seeks help from the president of DuPont, Crown's chemical supplier.

1951 *I Love Lucy* debuts on American television.

1953 Crown celebrates 75 years in business with a book penned by Clifford H. Anderson.

1954 Dr. Jonas Salk develops the polio vaccine.

1957 The U. S. Department of Agriculture reports that American farmers have more acreage plants in soybeans than ever before and soybean exports reach 80 million tons.

1957 U.S.S.R. launches Sputnik I and II, the first space satellites.

1960s To boost revenues, Crown pursues more structural steel business and wins a contract for the outfield fence at the Twin Cities' new Metropolitan Stadium, steel fabrication for the first skyway built over Marquette Avenue in downtown Minneapolis and an all-steel stage for the Minneapolis Symphony Orchestra.

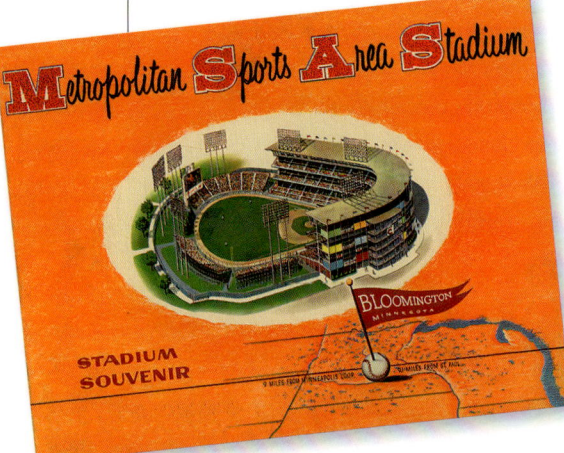

1961 John F. Kennedy becomes America's 35th president; he is assassinated in Dallas two years later.

Crown employees 1962

1963 Crown closes its foundry, dating back to its earliest years, and its ornamental iron department. Four departments remain: structural steel, extraction, feed screws and fence sales.

1964 Martin Luther King, Jr., wins the Nobel Peace Prize; he is assassinated four years later.

1965 Clifford I. Anderson, representing the third generation of Andersons at Crown, joins the company. His brother George becomes a full-time engineer in 1969; their brother Austin becomes a director in 1981.

1969 Golda Meir is named prime minister of Israel.

1970s Crown's annual sales grow from $4.2 million to $15 million by the decade's end.

1970s Crown builds 13 of the Twin Cities' unique skyway bridges and, in this decade, fabricates structural steel for projects such as the IDS Crystal Court, new buildings on the University of Minnesota campus, additions to Sister Kenny Institute and Abbott-Northwestern Hospital, Lutheran Brotherhood corporate headquarters and the Science Museum of Minnesota.

During the 1970s, Crown fabricated structural steel for many Twin Cities projects including a new Coon Rapids High School.

1970 In its first overseas partnership, Crown and the Chajuss family of Ashdod, Israel, begin a multi-decade collaboration to produce high-quality Soy Protein Concentrate (SPC).

1971 The 26th Amendment to the U.S. Constitution grants 18-year-olds the right to vote.

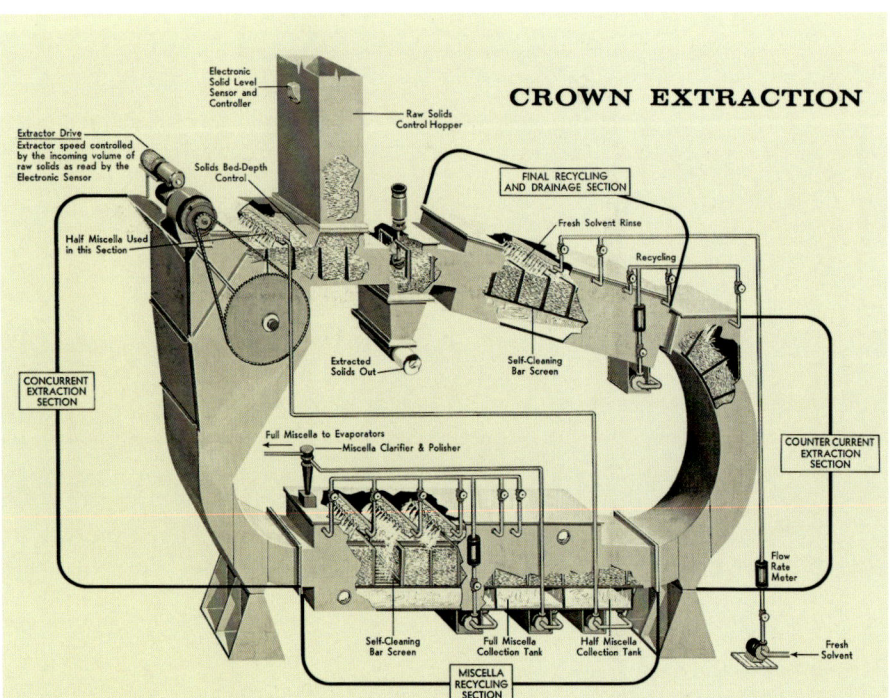

CROWN EXTRACTION

Electronic Solid Level Sensor and Controller

Raw Solids Control Hopper

Extractor Drive Extractor speed controlled by the incoming volume of raw solids as read by the Electronic Sensor

Solids Bed-Depth Control

FINAL RECYCLING AND DRAINAGE SECTION

Fresh Solvent Rinse

Half Miscella Used in this Section

Recycling

CONCURRENT EXTRACTION SECTION

Extracted Solids Out

Self-Cleaning Bar Screen

Full Miscella to Evaporators
Miscella Clarifier & Polisher

COUNTER CURRENT EXTRACTION SECTION

Self-Cleaning Bar Screen

Full Miscella Collection Tank

Half Miscella Collection Tank

MISCELLA RECYCLING SECTION

Flow Rate Meter

Fresh Solvent

1974 Crown wins its largest structural steel contract involving fabrication as part of a $25 million coal handling facility on Lake Superior.

1975 A major slump in business forces Crown to lay off 10 percent of its plant staff.

1977 Crown sells its first Heinz Schumacher invention—Desolventizer-Toaster-Dryer-Cooler (DTDC), operating on peanut cake—for Sessions Oils Mill in Alabama. (It is still operating today.)

1977 Jimmy Carter becomes the 39th U.S. president.

1971 Crown lands its first contract with giant Cargill—a 700 tons-per-day soybean extractor in Cedar Rapids, Iowa.

1972 Arab terrorists kill two Israeli Olympic athletes and nine other hostages in Munich.

1972 George Anderson takes a lengthy trip to Taiwan to assist with building and start-up of a new soybean processing plant.

Clifford I. Anderson

1973 Crown posts its first $1 million sale in the solvent extraction business with sale of a 2,000 tons-per-day plant to Farmers Grain Dealers Association in Sergeant Bluff, Iowa.

1974 Clifford I. Anderson becomes Crown's new president.

1974 President Richard Nixon resigns after the Watergate scandal; Vice President Gerald Ford becomes America's 38th president.

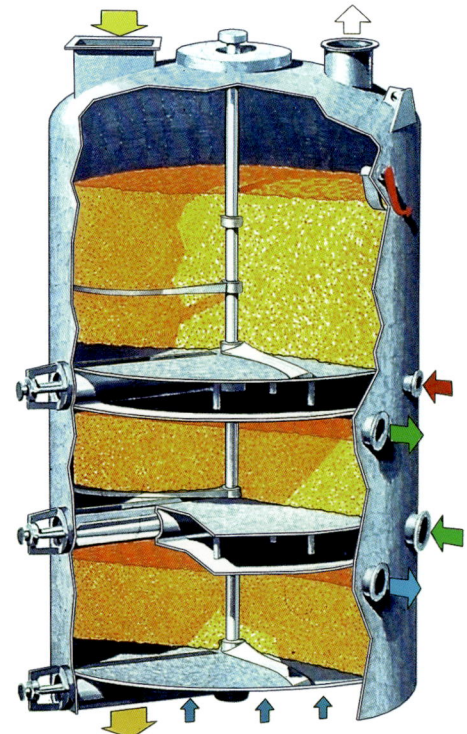

Crown's original Schumacher DTDC

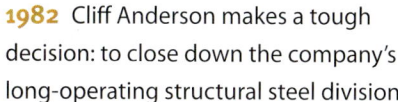

1978 Crown celebrates its Centennial.

1978 The first test-tube baby is born in England.

1979 Crown wins the confidence of another of the "bigs" in agribusiness: Archer Daniels Midland (ADM). They order a 1,200 tons-per-day plant with a stainless steel extractor and DTDC.

1979 Glenn Brueske and George Anderson take their first trip to China.

1979 The Shah of Iran is forced into exile and is replaced by Ayatollah Khomeini.

1980 Ronald Reagan is elected the 40th U.S. president.

1980 Crown sells its largest soybean processing plant to date, a 2,500 tons-per-day plant for Boone Valley Cooperative in Eagle Grove, Iowa.

1981 Crown buys its first personal computer and introduces engineering on spreadsheets.

1981 A licensing arrangement with British company Rosedowns opens the door for Crown's sales to customers outside the United States, especially in Europe.

1982 The first Crown-Schumacher Counterflow DT is put into service, continuing the highly successful collaboration with German engineer Heinz Schumacher.

1982 The Vietnam Veterans Memorial is dedicated in Washington, D.C., with the names of more than 58,000 fallen soldiers inscribed in granite.

1982 Cliff Anderson makes a tough decision: to close down the company's long-operating structural steel division.

1983 Along with the venerable manufacturer Krupp, Crown is retained to help rebuild Oil Mill Hamburg's solvent extraction plant in Germany.

1983 The U.S. space shuttle Challenger is launched on its maiden flight. Three years later it explodes, killing seven crew members.

1984 A silicon microchip is introduced that stores four times more data than previously possible.

1984 Crown Holdings, Inc., is created to serve as the parent company of Crown Iron Works' former divisions.

1984 Crown has its equipment operating in countries including Argentina, Australia, Bolivia, Brazil, Canada, Denmark, France, Germany, Great Britain, Israel, Jamaica, Mexico and Taiwan.

1984 Crown's old "iron works" heritage continues as Crown Auger Manufacturing, Inc., of Cokato, Minnesota.

1984 A powerful Iranian family hires Crown to build a small pistachio extraction plant in that country.

1986 The world's worst nuclear accident occurs when a reactor blows up at Chernobyl power station near Kiev, U.S.S.R.

In 1980 Crown wins a $1.8 million contract to provide all structural steel for the Hubert H. Humphrey Metrodome in Minneapolis.

1986 A new corporate home for Crown Iron Works and Crown Holdings is established at 1600 Broadway Street Northeast, Minneapolis.

1987 Crown sells Crown of Minnesota, Inc., purveyor of mink ranching equipment, pet cages, fencing and playground equipment.

1988 After years of patience, Crown sells its first plant in mainland China—a cotton-seed operation ordered by Cargill in Jinan.

1988 To expand its business in edible oils and oleochemicals, Crown purchases Wurster & Sanger of Chicago.

1988 Crown receives its largest spare parts order in history (for a 3,000 Series Crown Extractor) and sells its largest DTDC to Lauhoff Grain Company of Danville, Illinois.

1989 Competitor DeSmet of Belgium buys Rosedowns, thus ending Crown's working partnership with Rosedowns.

1989 The Berlin Wall is opened by East Germany and later torn down.

1989 George H. W. Bush becomes America's 41st president.

1989 Crown's first Model IV Extractor debuts at a specialty extraction plant operating by Schering Corporation in Mexico.

1990s Pioneering work in soy protein concentrate (SPC) positions Crown as a major supplier of equipment to produce SPC worldwide. Crown explores new ventures in specialty extraction on many fronts and creates a division to focus on this work in 1997.

1990 A group of former Rosedowns engineers who, a year earlier, had formed Oil and Fats Engineering Ltd., join forces with Crown Iron Works and become Europa Crown Limited. Crown also forges a joint venture with Oiltek Sdn Berhad of Kuala Lumpur, Malaysia, and ChemTech of Britain.

1990 The first edition of Microsoft Windows 3.0 software is shipped to consumers.

1990 Crown sells its new dehulling system to Rose Acre Farms in Seymour, Indiana—the first of many contracts for the new process.

1990 Crown's oilseed business delivers $16 million of the company's total $18 million in annual revenues. Crown earns the government's "E" Award for excellence in American export activity.

1990 Crown and Kin Kong Yee of Pisces Engineering, a manufacturing company in Malaysia, inaugurate a working relationship that opens new markets.

1991 More growth for Crown triggers a move to new quarters at 2500 West County Road C in Roseville, Minnesota.

1991 Operation Desert Storm begins in response to the Iraqi army seizing Kuwait.

1992 The SAO project in Bolivia represents Crown's first turnkey project in South America.

1992 10,000 cellular phones are sold in the United States.

A 4,000 Series Crown Extractor

Matl flow			1.08	1.08		
Tray OUT	180	180	260	260		
CHUTES/tray	1	1	2	1		
Chute W"	15.25	15.25	15.25	15.25		
Chute R"	43.5	43.5	84	84		
	13.1	13.1	9.7	9.7		
W up "elev. S"	0	0	28	0		

2002

TRAY

This program calculates estimates of the CIRC or the capacity of the tray to circulate me... ...tray to rea... the next chute. This is in raw seed...

1993 Crown sells its 151st extractor, a unit with a capacity of 3,000 tons-per-day.

1993 As revenues increase, Crown's Spare Parts Department is elevated to the Renewal Parts Division.

1993 Crown enters Brazil with a 1,500 tons-per-day soybean processing plant for Coopersul and begins its active working relationship with Intecnial.

1993 William J. Clinton becomes the 42nd U.S. president.

1995 Expanding services to Central America, Crown opens an office in San Pedro Sula, Honduras. Crown's Moscow office opens the same year.

1995 While serving on Crown's Board of Directors, Clifford H. Anderson dies at age 91.

1996 Crown's first Diflow Deodorizer is built for processing edible oils with high efficiency and minimal energy consumption.

1996 AT&T introduces Internet access service.

1998 The Dow Jones Industrial average hits 9,000 for the first time in a single day's trading.

2000 By the twenty-first century, international business represents approximately 80 percent of Crown's annual revenues.

2000 Crown sells its first complete preparation and extraction plant in China to East Ocean Oils of China.

2000 Crown begins research into supercritical CO_2, a liquid that acts as a gas that can replace solvents in oil extraction from oilseeds. This work leads to a patented process and registered trade name, HIPLEX®.

2001 Three hijacked commercial jetliners destroy the World Trade Center towers in New York City and hit the U.S. Pentagon, killing approximately 3,000 people. A fourth hijacked plane crashes in a field in Pennsylvania.

2001 Crown forms a joint venture with FEC, Friendship Engineering Company of Wuhan, China, naming it Crown Friendship Engineering Company.

2001 West Central Cooperative hires Crown to help design a larger and better biodiesel plant in western Iowa. The plant opens in 2002 producing 12 million gallons yearly.

A Crown Friendship Engineering Company plant in Wuhan, China.

2002 West Central Cooperative and Todd & Sargent builders form an alliance called Renewable Energy Group (REG). They name Crown as their chief process technology provider.

2002 Crown is among 60 Minnesota companies accompanying Governor Jesse Ventura on a trade mission to China.

2002 Twelve European nations start using the Euro, a common unit of monetary exchange.

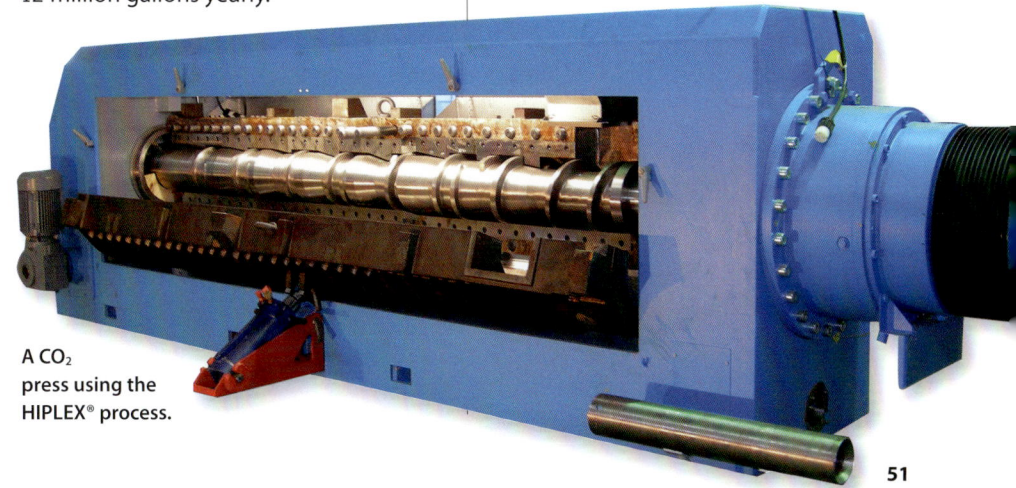

A CO_2 press using the HIPLEX® process.

2003 Crown invests in a new solvent extraction lab that features a continuous pilot-scale crushing facility to develop new applications and test proprietary processes.

2004 Crown announces a joint venture with Kumar Metal Industries, a company based in Mumbai, India.

2004 The U.S. government offers subsidies to stimulate biodiesel production.

2004 George W. Bush is re-elected president for his second term.

2004 Crown sells CAM Manufacturing, Inc., of Cokato, Minnesota.

2004 Crown is named one of Minnesota's top 50 "fastest growing private companies" by *Minneapolis/St. Paul Business Journal*. They earn the distinction again in 2005.

2005 Clifford Anderson joins Minnesota's Governor, Tim Pawlenty, on a second trade mission to mainland China.

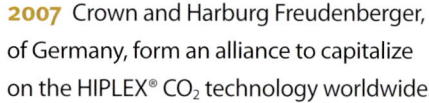

Crown employees celebrate their 125th anniversary in 2003.

(Above): Cliff Anderson (left) and George Anderson (right) celebrate Harold Anderson's earlier career with Crown Companies in 2003. (Right): Clifford H. Anderson (seated) is pictured with Harold Anderson seven decades earlier in 1938.

2006 By September 30, *Biodiesel Magazine* reports U.S. plants produced 250 million gallons of biodiesel in fiscal 2006; 77 million gallons come from Crown plants.

2006 The Internet and computer-aided design allow Crown project managers to oversee plant installations continents away.

2006 The oldest of the baby boomers—those born between 1946 and 1964—turn 60.

2007 Crown and Harburg Freudenberger, of Germany, form an alliance to capitalize on the HIPLEX® CO_2 technology worldwide.

2007 Interstate 35W in Minnesota collapses into the Mississippi River on August 1, causing injury and death and prompting widespread scrutiny of the nation's bridges.

2007 Iowa State University, Crown and Safesoy—an Ellsworth, Iowa-based company—start up a plant designed to use CO_2 technology in production of edible protein, oils and fuel from soybeans.

2008 Crown has sold 21 biodiesel plants, six of which are located outside the United States.

2008 Crown Companies celebrate 130 years in business. Clifford I. Anderson, president for 34 years, retires.

Crown Joins the "Global Village"

"Your company has experienced quite an interesting and exceptional year . . . with challenges galore." The elder Clifford Anderson said it succinctly when he described 1982, the year Crown Iron Works closed its Structural Steel Division after about 80 years in the business. He credited his son, Clifford I., and his management team with "doing a very difficult job under difficult circumstances. . . . it is not easy to cut back, meaning that many good people had to be told they no longer had a job with us."  But the elder Anderson was optimistic.

A 700 tons-per-day Crown extractor is unloaded at the docks in Hull, England, bound for Rostov-on-Don, Russia. A welder secures a tail section of a Crown extractor.

(Above, Left to Right): Crown's new office and warehouse on Broadway and Johnson Streets in Northeast Minneapolis, circa 1986. George Anderson was an early adopter of personal computing. By 1981, he used a Radio Shack computer with a spreadsheet to do major engineering work.

> "If all goes well, we should have a good, solid-looking 2-story building with 7,500 feet on each floor. But we can't afford to build as fancy a building as we had when we remodeled our old building back in 1946!"

Clifford H. Anderson, Crown's chairman, describing Crown's new office and warehouse

Republican to his bones, the Chairman cheered President Ronald Reagan's plan to heal the nation's economy: "If he doesn't succeed," Anderson warned Crown employees, "the country will go back to the spend and tax policy that caused the present unhappy recession." Unconcerned about political correctness, Anderson trained his eyes on his listeners, "We should pray for him to succeed."

The New Crown, Circa 1984

Crown went to work selling or scrapping equipment left idle and inventory remaining after closing its Structural Steel Division based in the company's aging Tyler Street facility. Income from these sales plus work in process and structural steel receivables actually left Crown with no bank debt and about $1.4 million invested. Not bad for a company that eliminated one of its historically largest divisions.

Shutting down structural wasn't the only big change in Crown's life in the early 1980s. True to its Northeast Minneapolis roots, when Crown sold its Tyler Street plant, it moved only blocks away to stay in the old neighborhood. Cliff Anderson and his management team created a new company, called Crown Holdings, Inc., in 1984 to serve as the parent company for the former Crown Iron Works divisions—each with its own management. The new snapshot of Crown looked like this:

Crown Iron Works Company, moving to a new office and warehouse on Broadway and Johnson Streets in Northeast Minneapolis, would focus on design and engineering of solvent extraction processes and plant installations for the oilseed, food and chemical industries. By 1984, Crown had international licenses and manufacturing contracts with customers in

Argentina, Australia, Bolivia, Brazil, Canada, Denmark, France, Germany, Great Britain, Israel, Jamaica, Mexico and Taiwan. In the U.S. market, perhaps 75 percent of new oilseed solvent extraction equipment being sold carried Crown's logo.

"When we got out of structural steel, we were booming in soybean processing," Cliff Anderson remembers. "We were so busy that we had to farm out work to other companies in order to meet customer needs. Before long, we realized we were doing just as well on that work as we would making the equipment ourselves."

In a decision that predated popular outsourcing practices by more than a decade, Crown began buying a majority of its custom-made equipment from a network of tested and trusted suppliers. "The decision to outsource was significant," says Gary Pulis, who joined Crown in 1978, a chemical engineer and plant superintendent from ADM. "It eventually led to fabrication in Europe, Malaysia, China, Brazil and India. The first job we did outside Crown was with Plant Maintenance Service Corporation of Memphis, Tennessee, who built three cooker-conditioners for Boone Valley Cooperative in Eagle Grove, Iowa. It was a different way of doing business and probably one of the best decisions Crown ever made." Along with Plant Maintenance, Whirl Air Flow of Minneapolis was among the first key suppliers.

Crown's old "iron works" heritage survived as **Crown Auger Mfg., Inc.,** a new company established in Cokato, Minnesota. This was Crown's only in-house manufacturing division, after the structural division disappeared. Although Crown relied heavily on outsourcing, Crown Auger would still supply about 20 percent of the products that Crown sold, including specialized conveyor screws and all the replaceable, proprietary moving parts for solvent extraction equipment.

(Above, Left to Right): Crown's processes are designed for many types of oilseeds including soybeans and sunflower seeds. The seeds are shown here in their original state (top) and processed state (below). This large Archimedes Screw Pump, produced by Crown, is used for sewage plants and flood control. At Hercules Corporation in Brunswick, Georgia, the stainless steel tail section of a 1,500 tons-per-day extractor is lifted into place.

"The augers produced by Crown, from ¾-inch in diameter to the huge 12-foot size, have many uses by a wide variety of industries. . . . grain handling, snow blowers, food processing, even sewage disposal systems."

The Cokato Enterprise,
November 14, 1984

> "From 1982 to 1984 Crown removed itself from some dying industries, improved its penetration of others and removed a lot of risk factors that are connected with heavy investments in manufacturing."
>
> **Clifford I. Anderson, president**

> "Our cash position is very strong and Midland Bank is giving consideration to increasing our $1 million credit line to $1.8 million, if necessary."
>
> **Clifford I. Anderson at the March 6, 1986, stockholders meeting**

Crown's third division, **Crown of Minnesota, Inc.,** on Central Avenue in Northeast Minneapolis, continued selling its mink ranching supplies, fencing, pet cages, playground equipment, gazebos, and a new addition—wood stoves. As Crown entered new markets, there was less time to focus on this division. In 1987, Crown sold Crown of Minnesota to its former managers, Stan Yokum and Joe Polnazsek.

In addition, Clifford I. Anderson and his brother George formed **C & GA Partners** to build Crown's new office and warehouse building.

The company's transformation was significant enough to issue a statement describing Crown Holdings, in October 1985. "Much of what people remember Crown Iron for in the last 107 years of its existence has been eliminated in favor of businesses that offer high cash flow, faster inventory turnover and more consistent margins," the company said. "Thus, Crown is no longer in foundry work, ornamental iron, forgings, curtain wall or window walls or structural steel." The transition started many years earlier, but the sale of Crown's old Tyler Street factory officially marked that chapter closing.

The Man Behind the Anvil

In autumn of 1986, when Crown greeted guests to the new Crown Iron Works/Crown Holdings building at 1600 Broadway Street Northeast, the company's ties to its history and the neighborhood were evident. A display included a pattern for old lamp posts Crown made to grace Nicollet Avenue in downtown Minneapolis, foundry tools dating back decades, a grave marker salesman's sample case, ornamental pieces from the 1920s, military barbed wire entanglement posts, and a pattern board for Ford convertible top bronze latch pegs, fashioned by Crown.

Models of Crown's soybean extractor solvent plant—representing the company's biggest business unit—were positioned front and center in the new lobby. At the front door, a huge, heavy blacksmith's anvil greeted visitors. It was the "workstation" for Lester Hefner, a Crown employee for 35 years who had died one year earlier. "The anvil is a reminder of a slender man with large, blacksmithing hands," a tribute to Hefner read. "He tackled a physically hard job daily and he was proud of the job he did. He was a dedicated, hard worker . . . faithful even when he wasn't feeling well. He felt the responsibility to his job and to Crown."

The Globe Trotting Begins

Crown was ebullient about its prospects in the 1980s. "We have never seen so many opportunities for acquisition as we do now," Cliff Anderson told shareholders. Crown looked outside the United States with equal enthusiasm: "We are exploring a stronger marketing presence in Mexico," he said. "Even though the Mexican economy is in bad shape, their government has declared that basic food processing is a national priority." Crown engaged Ralph Romero of R&D Equipment Company in Texas to represent the company. During the 1980s, Crown won its first order for a soybean plant near Mexico City and sold its first Model IV Extractor—a product of the company's research and development efforts—for a specialty extraction plant operated by Schering Corporation in Mexico.

Inflation Beyond Imagination

Soybeans had also become an important crop in South America in the 1980s. The climate and soil conditions were conducive to soybean farming and soybean meal was a preferred animal feed as poultry and hog production grew. While soybean oil had been merely a byproduct of processing, the South Americans soon saw the value of refined oil for human consumption. Cargill was already in South America, with Bunge and ADM not far behind. DeSmet was the major engineering and processing equipment resource, but Crown was making inroads.

In 1983, Crown had begun to stake out territory in Argentina when Glenn Brueske sold an extractor to Cargill and a Crown team including George Anderson, John Chasteen, Dean Nordquist and Bob DeLeo began traveling there to supervise construction and installation. "We usually traveled alone to Argentina and met up with our agent in Buenos Aires, Hector Sampedro, president of Howe-Baker of Argentina," says Anderson. "Most of the fabrication and installation work was done in association with Mario Allocco, of the manufacturing company bearing his name."

The United States was not a favorite in Argentina when Anderson first traveled there. The brief, but highly publicized, Falklands War had just ended and the United States had been suspected of helping the British against Argentina. "But," says Anderson, "they were generous enough never to mention that in our business dealings."

There were other obstacles. During the 1980s, business in Argentina was burdened with high import taxes between nations, it was nearly impossible to transfer money outside of Argentina, negotiations in Spanish were tricky for the Americans and high inflation was onerous.

"By the late 1980s, the inflation rate in Argentina averaged 5,000 percent," says Anderson, "and from March 1989 to March 1990 it was 20,000 percent. There was one period when Argentina's currency dropped an average of 25 percent in value each day for 17 days straight!" Anderson saw scenes of unrest when banks refused to release funds to customers and there was an ever-active black market for U.S. dollars.

"After a Crown job was quoted," Anderson says, "months later it was billed at a far higher price according to a formula for materials and labor values published each day. Bank interest roughly matched inflation and, in theory, everything worked fairly well. However, any payment delay was a disaster. When we billed a job, the payment was hand-delivered within a half day. Often, we presented a second, smaller bill to cover the inflation of the few hours it took the main payment to reach us and be deposited."

Profits for Crown in South America from 1983 to 1989 were razor thin, but the company sold several plants to Cargill, Buyatti and Vicentin. On the Vicentin project, Heinz Schumacher was able to build a new kind of soybean expander discharge system and test the effects of rapid cooking on protein.

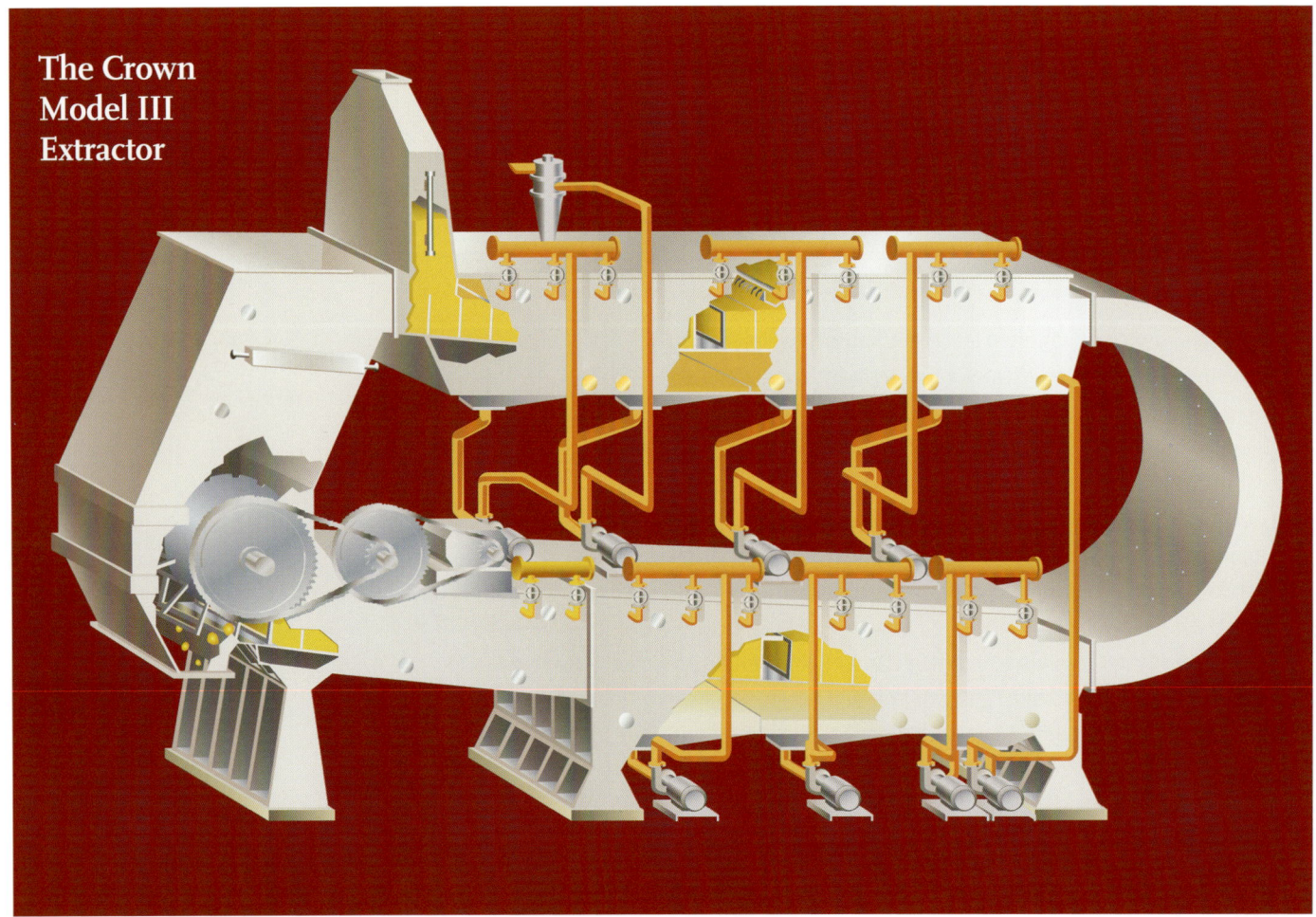

**The Crown
Model III
Extractor**

(Above): Crown's George Anderson drew the first sketch of this Model III Extractor on a restaurant napkin.

"Between Crown's extractor and the Schumacher DTDC, they became a real powerhouse."

Lou Kingsbaker, founder of C. L. Kingsbaker, Inc., oil seed industry consultants

Extractors in Europe

Looking across the Atlantic, Crown already had an exclusive agreement to sell English-made Rosedowns screw presses in the United States and Canada, a product that Crown believed was superior to the U.S. competitors' at the time. In addition, Crown had a licensing agreement with Rosedowns, a subsidiary of the British Simon group, giving them the right to sell and manufacture Crown's extractors and the new Heinz Schumacher designs in any country outside the United States. Rosedowns, the 200-plus-year-old company, had a solid reputation in Europe and more than 90 percent of its focus was on export.

Consorting With Krupp

Crown had its first introduction to the giant German manufacturer Krupp in 1983 when Oil Mill Hamburg's solvent extraction plant had a major explosion and fire and the company retained both Crown and Krupp to help rebuild it. Crown supplied two of its largest extractors, working alongside staffers from the legendary German icon of engineering.

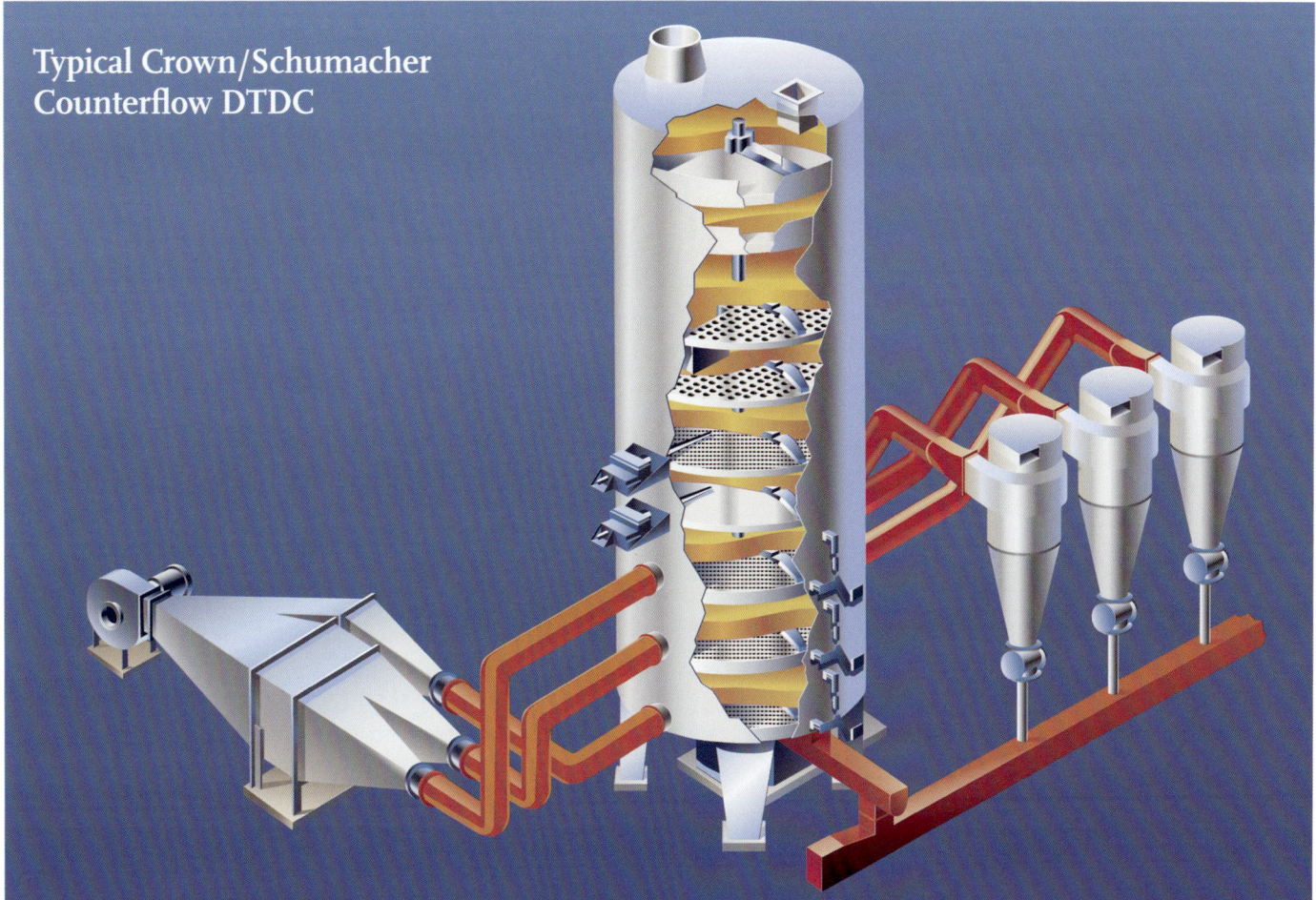

Typical Crown/Schumacher
Counterflow DTDC

"I will never forget Heinz Thiem, the Oil Mill's senior engineer, who was a young man during World War II and remembered a famous warship, the Bismarck, being built in Hamburg harbor," says George Anderson. "I visited him at his plant in the late '70s and '80s along with the inventor, Heinz Schumacher. Just for fun, Thiem used to call me 'boy' all the time and he said, 'I'll never buy your crude American machines.'" After the explosion wrecked Thiem's plant, he called Anderson and Crown's sales manager, Glenn Brueske: "Can you please come over quickly?" he said. "I want to talk to you about your extractors."

After the project was agreed upon by Crown and Krupp, Thiem took Anderson and Brueske aside. "He told us not to discuss the contract price," Anderson said, "because Oil Mill Hamburg paid a bit more for our extractors than Krupp's." (Many years later, Crown and Krupp would become industry allies.)

Geopolitical Pistachios

Opportunities in many countries surfaced in the 1980s: in Australia, Nigeria, Argentina, Venezuela, Brazil, Greece, Canada, Denmark, France, Holland, the Philippines, Thailand—and Iran. Crown

"Although we have sold some equipment for overseas use over the years, we have never been export-oriented. Now, as we try to make it our major market, we find it is a great learning experience."

Crown Iron Works division report in the Spring 1987 issue of *Crown Gazette*

(Above, Left to Right): By the late 1980s, Crown plants were computerized. This photo shows the control panel of a Hercules Corporation plant in Brunswick, Georgia. Jeff Scott and Augusto Skrzypek are pictured at an installation of a Series 1,600 extractor for Caramuru, Brazil. Crown's acquisition of Wurster & Sanger, Inc., in 1988 gave the company expertise beyond its primary business: edible oil extraction. Pictured here, Wurster & Sanger technicians start up an edible oil refinery in the Philippines. Sections of a Crown DT (desolventizer-toaster) travel a highway in England on their way to customers in Europe.

won a $580,000 contract for a small, 20 tons-per-day Iranian pistachio extraction plant in October 1984. Relations between the United States and Iran had been chilly for several years, and this sale was made possible when both nations briefly improved trade relations. The customer was a wealthy Iranian family with ties to senior leadership in the government. "America was limiting trade with Iran, but we managed to get a license to sell our plant because it involved food, not arms," Cliff Anderson remembers. "Iran welcomed foreign business in those days. We secured a letter of credit with a foreign bank and arranged for an inspector to examine our finished equipment at Plant Maintenance in Tennessee."

Thinking the order had passed inspection, Plant Maintenance loaded and shipped the equipment by truck and then had it transferred for shipping across the Atlantic. Meanwhile, the inspector informed Crown that the equipment did not pass some detail of inspection. The shipment left the United States but the shipping papers were incorrect and Crown's letter of credit evaporated. While at sea, the shipment was diverted to Turkey. Thinking they had time to rectify the problem while the shipment sat in Turkish customs, Crown breathed a little easier. "The shipment went through the port like greased lightning," says Anderson. Indeed it did. While the shipment was being held for inspection, it was pulled out of the fully-secured holding area and disappeared before the inspection took place. Without approval, the Iranian customer's agents had "hijacked" the shipment without telling Turkish customs or Crown.

When the shipment reached the Iranian border, border guards called Crown because its name was on the goods. George Anderson remembers a beautiful Saturday morning at the home of Jeff Scott when they negotiated by telephone with the U.S. branch president of Iran's Bank Saderat.

"Jeff worked out a deal," Anderson remembers. "They paid us in full, but we gave them a letter of credit to take back any amount up to half the total value of the shipment if they weren't satisfied when they opened the shipping crates. Apparently, because of their poor international relations and lack of access to legitimate shipping resources and inspections, they'd been burned on shipments of poor quality or empty crates. They wanted some recourse if we had cheated them. They got their shipment, they were satisfied and they never took back any money. By that time, the U.S. had again closed the doors on future trade."

The equipment was ready for installation about the time American-Iranian relations reached the breaking point. "When it came time to commission the plant, our customer asked us to send our key people over, but no Americans were allowed to travel to Iran," Cliff Anderson recalls. "Rosedowns volunteered to send some of their people, but at the very same time a group of Iranians made international news by kidnapping Terry Waite, a British national."

Ultimately, the customer, a chemical engineer, told Crown he would commission the plant himself. It was a time-consuming lesson for Crown and not the only international hotspot the company would encounter.

In China, Patience

After practicing years of patience, in 1988 Crown also sold its first plants in China—a cotton-seed operation ordered by Cargill for Jinan, China, just south of Beijing, and a sunflower/linseed plant for CNTIC, a Chinese company, in Zhangjiakou, a goodly distance away to the northeast. Crown was on a steep learning curve in that culture, especially when it came to selling local customers: "It appears that everything in China takes a lot more time than anticipated," one

"A lot of our Crown Iron jobs are foreign now—in Europe and Central America. Lately, the weaker dollar has helped us get more orders and more work for U.S. suppliers. We aren't fussy; we just give the customer the best all-around package of price, quality, reliability, delivery."

Cliff Anderson, in January 1987

> "Cargill came in with a 'shopping list' for the year. They have ordered an extractor, two DTs and two DCs to be built throughout 1988."
>
> ***Crown Gazette* divisional notes**

> "The Wurster & Sanger Division has been successful in obtaining new orders from Malaysia and Thailand for margarine plants and they've closed an order for a complete oil refining plant for El Salvador."
>
> ***Crown Gazette*, Spring 1989 issue**

divisional report said. "China is a tougher market than we ever dreamed." Crown's efforts would finally pay off in the 1990s with multiple plants built in mainland China.

From Chicago, a Good Fit

When it acquired sixty-seven-year-old Wurster & Sanger, Inc., of Chicago in 1988, Crown was able to expand its expertise beyond its primary business: edible oil extraction. Founded in 1921, Wurster & Sanger's niche was supplying engineering and equipment for the edible oil and oleochemical industry. The company also had a strong suit in engineering equipment for margarine and glycerin plants—a specialty that would prove the most beneficial to Crown. "Wurster & Sanger's products are sold around the world, mostly to the same customer group that we seek for Crown's solvent extraction equipment," Cliff Anderson told Crown employees. "In return, Crown is strong in the U.S. and Canada where Wurster & Sanger has not been as dominant." It was a good fit, and although Crown discovered that some of the technology it acquired had to be updated, Wurster & Sanger opened up new opportunities around the world.

"When we bought Wurster & Sanger, there was a strong market for margarine processing plants in southeast Asia," says Jeff Scott, Crown's architect of international growth. "One of their greatest assets was Barry Smith, who was amazingly knowledgeable about margarine formulations. He would put a sample in his mouth, taste it, analyze it and write a formula for the sample. Our customers loved him because they could have their own proprietary margarine formula and brand."

A Breach and a Bond

Along with overseas contracts from longstanding American customers Cargill and ADM, the spark that ignited Crown's international business was Rosedowns of Hull, England. Crown's equipment enhanced Rosedowns' product line in the global market and the company had solid relationships with fabricators who could manufacture Crown's equipment overseas.

But all that changed in the late 1980s when Crown's biggest European competitor, DeSmet of Belgium, bought Rosedowns from its parent Simon.

Before long, a handful of Rosedowns' best engineers left their old company. "They had competed with DeSmet their whole life and they decided to go their own way," says Scott. The group was led by an ambitious entrepreneur, Graham Goforth. They named their company Oils and Fats Engineering Ltd., in 1989. "At first, they were selling smoked salmon and rice, anything to make a buck," Scott says. "They were importing and exporting products."

At an American Oil Chemists' Society conference, former Rosedowns engineer Allen Forster approached Glenn Brueske and Scott. "They proposed focusing on the European market and representing Crown, just as Rosedowns had," he says. After much discussion, the leadership at Crown decided to join forces with the young company. Though the Brits were new to running a venture, they were old hands in the oilseed processing business and savvy engineers with solid international connections.

One year later, in 1990, Oils and Fats Engineering would join Crown Iron Works in a venture with a new name, Europa Crown Limited. "Shortly after we formed the company, ADM swept

A Handshake Is a Deal

Call it the Midwestern work ethic. Label it "Minnesota Nice." Crown Iron Works is known for fairness in business. "We work with our customers, our suppliers and our inventors fundamentally the same way," says Cliff Anderson. "We collaborate with them and we don't stiff them."

Adversarial relationships can quickly develop when contracts are valued in the millions and innovative ideas have patent and commercialization potential. Negotiations can get testy, but, says Anderson, "we mostly get past that, and we have trouble with people who can't. If someone is hard-nosed or trying to take advantage of us, we get fed up. We're not used to that kind of behavior."

Taking the high road has been Crown's policy in business. It has paid off in longstanding relationships and customers who return again and again.

Crown's collaboration with Iowa State University dates back to the 1940s when the company commercialized the "Iowa State College process," leading to a revolutionary soybean extraction method. It has continued with sharing ideas, equipment, knowledge and experiments.

"I always thought Crown was at the forefront of innovation. There weren't many companies innovating in the industry, but Crown always was. They were generous in sharing their knowledge with us."

Dr. Larry Johnson, who knew Crown at Texas A&M before he joined Iowa State to run its Center for Crop Utilization Research.

Riceland Foods—a $1 billion, farmer-owned cooperative founded nearly 100 years ago in Stuttgart, Arkansas—is the largest rice miller in the United States. The cooperative expanded from processing rice to soybean processing and rice bran oil extraction. "I began working with Crown when I joined Riceland in 1988 and I found them to be very solid, very dependable people," says Leo Gingras, a vice president of Riceland. Gingras says Crown's shallow bed extractor design is easier to maintain and operate. In addition, Crown's development of a deodorizer for vegetable oils helped Riceland produce a higher quality oil, retaining beneficial nutrients.

Crown's ingenuity in developing an extractor to produce rice bran oil has given Riceland Foods a premium product in high demand by large food processors.

"If they tell you they're going to do something, they do it, which in today's environment is a pretty rare commodity. They always follow through. They do great engineering and they make fine equipment."

Leo Gingras, a vice president of Riceland Foods

For years, Crown has engineered and built multimillion-dollar soybean extraction and DTDC equipment for Cenex Harvest States (CHS). A $13 billion, Minnesota-based company, CHS represents the merger of two, mega-farm cooperatives. Dennis Wendland, senior vice president of Oilseed Processing, has worked with Crown on CHS's major plants in Mankato and Fairmont, Minnesota.

"The extractor and DTDC are the two biggest pieces of equipment in a plant and reliability is crucial," says Wendland. "We have to be running 24/7, 365 days a year. This equipment has to meet our quality specifications and run continuously. If one goes down, the whole plant is shut down and every minute means money lost.

"Crown is the premier oilseed and solvent extraction processing company," says Wendland. "They rank first or second in the world, depending on who you talk to."

Hershel Austin was a Crown customer for 40 years. He recalls 1971 when Cargill bought its first Crown shallow bed extractor for its Cedar Rapids, Iowa, soybean processing plant. Several years later, Austin supervised installation of a Crown extractor in New South Wales. "When we were ready for the plant start-up George Anderson had recently married," says Austin. "He and his new bride came to Australia as almost a second honeymoon to help us."

Austin has supervised Cargill plants in the United States, Australia and Southeast Asia and has purchased Crown equipment for use around the world. When he retired, Austin was Vice President of Operations for Cargill's Oilseeds Group. "When I think of Crown, I think of integrity of the company, its management and its employees," says Austin. "I have always appreciated their simple and uncomplicated negotiations. With Crown, a handshake is a deal."

Crown's Europa Crown Limited office is located in Hessle, a suburb of Hull, England.

through Europe like Patton through the North Coast of Africa," says Scott. "They bought many independent operators and moved from being a minor player to a major player in oilseed processing in Europe. This gave Crown better name recognition across the pond because we worked with both ADM and Cargill." Even though DeSmet was the "800-pound gorilla" outside the United States, Crown and Europa Crown Limited began gaining market share.

Pairing With Pisces

That same year, Crown also formed a joint venture with Oiltek Sdn Berhad of Kuala Lumpur, Malaysia, to promote and sell Crown extractors there. In addition, Crown bought ChemTech, a British-based manufacturer of equipment used in making margarine. The working relationships with Oiltek and ChemTech were disappointing, but a strong bond with Pisces, a Malaysian-based equipment manufacturer, came from it. "Pisces was a young start-up company selling fabricating equipment to Oiltek and ChemTech in Malaysia," says Cliff Anderson. "The company's owner, Kin Kong Yee, was Chinese and our project for Malaysian Flour Mills was his first big job." It wasn't long before Pisces was manufacturing Crown's equipment for many locations, including Bolivia.

Jeff Scott outlined Crown's needs for Mr. Yee: "I told him we were looking for trustworthy fabricators and I said, 'We don't like to shop around and beat everybody up on price. We're looking for a partner.' He said, 'I'm your man.' They all say that when they think they're working with a big American company. K.K. just didn't realize how small we were at the time. Today, we share so many projects, we talk to K.K. almost daily.

Pisces Engineering was a young start-up company in 1990 when Crown began working with Kin Kong Yee, its owner. Since that time, Crown and Pisces have collaborated on many projects around the world.

"Our goal was to find joint venture partners with manufacturing capabilities, as well as sales and marketing skills," says Scott. "We wanted to build long-term relationships." To succeed, Crown and its joint venture partners had to be willing to adjust their prices so that everyone involved made a fair profit. "I think the philosophy worked out well," says Scott. "We found companies we could trust with capabilities we didn't have."

Home Runs for Crown

By the late 1980s, Crown hoped that its international growth would help offset a lackluster market at home. "Crown Iron Works is still feeling the effects of over capacity in the oilseed industry in the U.S.," the *Crown Gazette* told employees in spring of 1987. "The falling dollar has helped the processors, but not enough to look for new plants."

It seemed the only good news that fall in Minnesota was the Twins winning the World Series.

Within a year, however, the picture was better for Crown. The company won an order for Crown's largest DTDC for Lauhoff Grain Company in Danville, Illinois, and its largest spare parts order in history for a 3,000 Series Crown Extractor.

This reversal of fortune saw Crown's office almost empty ("Our people are seeing vendors and customers, working long hours and eating fast food," Cliff Anderson noted).

By year-end 1988, Crown had posted record sales and profits. "The swing of good fortune from a terrible 1987 was dramatic," Cliff Anderson told employees. Yes, there were backlogs, long hours and marathons of multitasking for Crown people in 1988, but Crown people pulled together to get the work out.

At holiday time, 1989, Crown had much to celebrate:

- Crown Iron Works had won more solvent extraction equipment orders than in any year in the company's history, including one for a new Model IV solvent extractor serving a brand new niche market: hard-to-extract materials,
- The Wurster & Sanger division shipped a record number of products,
- Crown Auger sold more earth augers than in any year in history—by a wide margin,
- Computer Aided Design (CAD) made its debut at Crown, and
- Overall, more than half of Crown's new orders were destined for locations outside the United States.

It was a challenging, expensive year for Crown, a time of growing pains, Cliff Anderson said in late 1989. But, he added: "We are building a solid base for years to come, in a world where some businesses are endangered to the point of extinction."

Indeed, America's business community was taking a beating. In 1989 alone, the Exxon Valdez dumped 11 million gallons of oil in pristine Alaska when its tanker ran aground. High-flying Michael Milken, a New York stockbroker, was indicted for fraud and whispers of more ethics lapses in the stock market surfaced. America's sinking savings and loan industry prompted President George H. W. Bush to avert total collapse with a $300 million bailout. And a cataclysmic stock market crash, two years earlier in October 1987, had pummeled Wall Street, companies and investors, triggering "downsizing," "rightsizing," and job losses.

Perhaps from previous "near death" experiences in its history, Crown had the perseverance and staying power to endure through all of it. 👑

An Old Company With New Tricks

While America's economy wheezed and staggered through 1990, Crown's solvent extraction business was robust, reporting "the biggest year ever in dollar volume"—twice the activity of 1988. "This means more than double the engineering," Crown's biggest division reported, "and we don't have double the people." But Crown's indomitable spirit was evident: "We CAN do it," the division promised in the *Crown Gazette*.

Pictured after the sales graph are: an internal and an external view of a Jet Dryer, soybeans shown at various stages in hot dehulling and, in the blue background, a stack of Jet Dryer rollers. Crown's dehulling process produces more than half of the world's high-protein/low-fiber meal.

> "We have been working on a new dehulling system for a long time. If this works, as it should, it will be a completely new product line."
>
> **Crown Gazette, Spring 1990 issue**

A glimpse inside a Crown Jet Dryer where soybeans go through their final heating and drying stage.

> "Our dehulling system was totally new, but Rose Acre Farms entrusted its first system to Crown."
>
> **Gregg Haider, Crown preparation sales manager**

Crown's extraction business delivered $16 million of the company's $18 million in annual revenues and its international business garnered recognition. Crown was given the "E" Award, created 30 years earlier during the Kennedy presidency—the nation's highest honor recognizing American exporters. Crown was in good company, joined by much bigger enterprises including fellow Minnesota company 3M and fast-growing Sun Microsystems.

The Dehulling Gamble

News that year of Crown's first contract for its new dehulling system for Rose Acre Farms in Seymour, Indiana, held promise, but there were obstacles. Crown had devised a brand new process—quite different from conventional dehulling methods. Crown needed to complete development and make final design and supply decisions to ensure an efficient and safe system. As Crown entered this new market it had competition from experienced players.

About five years before the Rose Acre Farms sale, Joe Givens (then retired) challenged Crown to devise an improved soybean dehulling system. Bill Stevenson, Bill Kratochwill, Irwin Lee Irwin, Darcy Moses, George Anderson and Joe Givens composed the initial hot dehulling development team and a 70 tons-per-day pilot plant was installed at Honeymead in Mankato, Minnesota.

Crown's entry into preparation of oilseeds proved to be a costly, five-year gamble, but after the initial sale to Rose Acre Farms in 1990, orders multiplied in the United States and especially overseas.

Gregg Haider, Crown's preparation sales manager, remembers the early years. "I interviewed with Bill Stevenson at Crown in 1992 and a couple days later, he offered me a position. He told me I would leave within days for Indiana for the start-up at Rose Acre Farms." Crown wanted Haider to work in research and development as well as using his Computer Aided Design (CAD) training to transfer paper drawings of Crown's dehulling designs to computer formats.

"I met Mike Cheney at an Indiana airport," says Haider. "He was researching Crown's new system for Honeymead in Mankato, Minnesota. One year later, in 1993, Honeymead bought Crown's second dehulling system." It was Crown's first, full-blown, four-line hot dehulling system.

Taking It on Faith

Inventors Darcy Moses (left) and Bill Stevenson work to fine-tune a prototype Jet Dryer.

Crown Iron Works operated for 70 years without any product differentiation until Clifford H. Anderson took a calculated risk, in 1947, and licensed an embryonic soybean extraction process. There were heavy losses and unexpected calamities, his son Cliff told members of the Process Equipment Manufacturers Association (PEMA) in 1990, but that decision led to decades of new products and applications. The ideas came from inventors outside the company and innovative thinkers inside Crown. Crown learned some important lessons about product innovation, Cliff Anderson told his PEMA colleagues.

"If you want to grow through product differentiation," Anderson told them, "Own your company. Taking this road is scary . . . and it requires a lot of faith. Don't have anybody to answer to."

Within reason, Anderson said, be generous to inventors. Treat them well and they will remember. "Your reputation will bring others to your door."

Collaborate with key customers in product development, Anderson said. "Develop a relationship with a customer who can test your ideas in his production facility." You'll have a real-world R&D site and your customers will have the benefit of your newest ideas.

Finally, Anderson said, be competent and strong enough to serve your customer. "Or," he warned, "he will actively encourage your more credible competition to take your place."

(Above, Left to Right): This chart shows the rapid growth in Crown's soybean dehulling contracts with the number of systems sold (in blue) and metric tons-per-day produced (in red) from 1995 to 2007. Crown's Jet Dryer introduced higher temperatures that removed the hulls from raw soybeans more easily when compared to conventional methods.

> "Some of Crown's customers asked us if there was a different— and better—way to handle dehulling and Crown went after it."
>
> **George Anderson**

Corey Paulson was a mechanical engineering intern at the Mankato, Minnesota, Honeymead plant when Crown installed its second hot dehulling system in 1993. "I got in on the ground floor," he says. "My first job was to write the operations manual for the hot dehulling process. I had to understand it so my instructions would make sense to others on paper. No one knew how much dehulling work we'd have in the early years." It wasn't long before Paulson focused solely on these new installations and he added pages to his passport reflecting increased international travel to South America and China. It was more than a boy from Mankato, Minnesota, could ever have imagined.

"I traveled to central Brazil and spent the night in what looked like old Army barracks with an air conditioner that sounded like an air compressor," he recalls. "In front of a hotel, it wasn't unusual to see a donkey cart next to a new BMW." Paulson spent five weeks with Crown project manager Noel Rosenthal running the new hot dehulling operation and teaching the customer's staff how to operate it. "I used a Portuguese dictionary, but mainly we relied on hand sketches," says Paulson, today Crown's technical manager for preparation and presses. "When discussing temperatures and pressures, a notepad and a pencil was a must."

Keep It Simple

The Crown system was simpler than the conventional dehulling process because it had fewer conveyers, motors and moving parts. It reduced the amount of plant space necessary and it consolidated the steps involved. With simplification came lower maintenance costs. In addition, Crown's process dramatically reduced dust emissions and energy consumption.

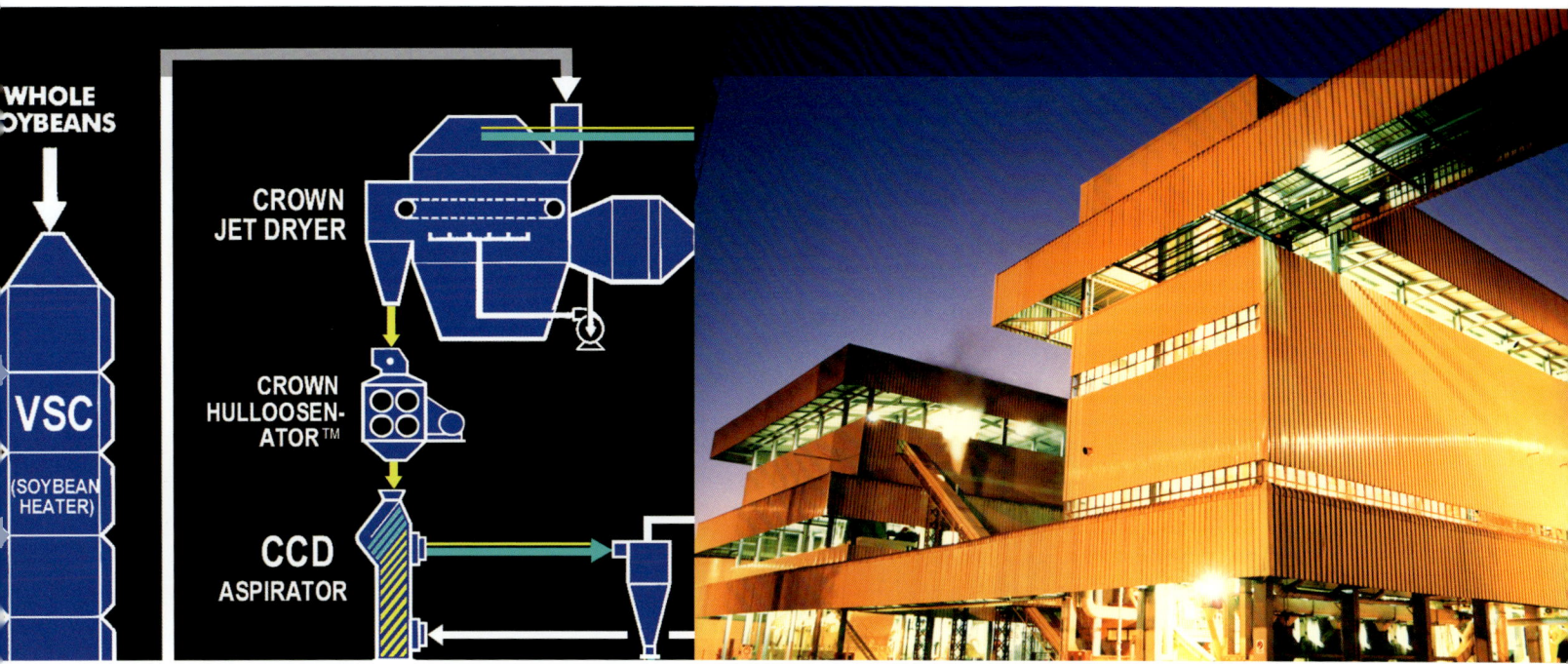

Crown's new system introduced higher temperatures, too. "In conventional dehulling, the maximum temperature was about 170 degrees while the beans dried. Then the beans sat in tempering silos for two to three days," says Haider. "Our system reduced that time to less than 45 minutes. In the first 30 minutes we raised the temperature of the beans to 140 degrees and then added a quick shock of heating, using a jet dryer, to bring them up to 190 degrees in two more minutes."

Treated this way, the hulls came off the beans more easily and their overall temperature and moisture were more consistent for the next steps: flaking and extraction. There was a greater risk of fire with increased heat, but Crown successfully addressed that issue. "The whole system not only improved dehulling, but it made the extraction process better because it produced a very consistent product," says Haider.

As orders came in, Crown kept tweaking its process. "We learned with each new installation," says Haider. "We added a patented aspiration feature to our bean heater that created better drying. We changed the whole air system that gave the operator control over each part of the process. We modified the jet dryer so it could also be used as a cooler." Over time, Crown discovered that it could sell the three main components of its dehulling process—the bean heater, aspirator and jet dryer—separately for different applications.

By the end of the 1990s, 25 Crown dehulling systems would be operating at plants in the United States, Brazil, Canada, China, Australia, Germany, Japan, Mexico and Argentina. Less than a decade later, in 2007, that number would spike to 71 plants, processing more than 40 percent of all the soybeans grown worldwide.

(Above, Left to Right): Crown's Vertical Seed Conditioner slowly heats soybeans to 140 degrees F. and dries them in only 30 minutes. Until this invention, conventional methods required two to three days in a "tempering silo." One of Crown's customers in Brazil hired an architect and turned his soybean processing plant into a show place.

"Several major processors are waiting to see how our system performs so they can decide whether to install ours or our competitors'."

Crown Gazette, **Spring 1992 issue**

Extraction Traction

As sales of Crown's new dehulling system ramped up in the early 1990s, the company's solvent extraction business also set records. By 1993, Crown sold its 151st extractor, a unit with a capacity of 3,000 tons-per-day, compared to 100 sold before 1985 (up to 1,500 tons-per-day) and 50 sold before 1974 (up to 1,000 tons-per-day). In addition, Crown's Wurster & Sanger group had so much success selling its oil processing technology to the edible oils and oleochemicals industry worldwide—it was elevated to division status.

Business observers credited Crown with its pioneering decision to outsource equipment manufacturing: "Today about 90 percent of Crown's extractor and processor components come from subcontractors, mostly in the South," the April 12, 1990, *Minneapolis Star Tribune* reported. "Because the processing plants and equipment are so large, they are assembled on site." With

> "It's our customers and our equipment that sell a new job for Crown."
>
> **Gregg Haider**

The Case for Growth

Always enamored with cars, Cliff Anderson used his recent visit to Morgan Motor Company of England to illustrate the need for growth and change in his own industry.

"Mr. Peter Morgan, the owner and grandson of the founder, hasn't modernized the company at all," Anderson said in his summer 1991 Crown Gazette *message to employees. "Most of Morgan's methods and the cars are nearly the same since at least 1935. They have a two-year backlog, make a profit and everybody seems content."*

"Why doesn't Crown relax and follow Morgan's example?" Anderson asked.

"Our product, unlike sports cars, does not get cute or collectible with age," he explained. "Machines are only valuable for what they can do, and technology changes make older designs worth less."

Second, Crown's major competition is "trying to crowd everyone else out of the marketplace," he said. "They would force us to sell out if we weren't aggressive."

Third, Crown's customers—even the oldest—also face changes in markets and products.

Fourth, the cost of doing business is not static, Anderson said. "It continues to rise with new demands for pollution control, health coverage for employees, and increased travel costs and taxes.

"A growing company generally improves the level of optimism and gives people a way to advance without just waiting for their leaders to retire," Anderson said.

The Morgan story is fascinating because it is so unusual, he concluded. "Only one automobile company in the world has done what they are doing. But for us at Crown, the odds of survival are not good if we try to stand pat."

Only six years after moving into new headquarters in Northeast Minneapolis, Crown outgrew them again in 1991 and settled into this building in Roseville, a suburb of Saint Paul, Minnesota.

exports accounting for more than half the company's oilseed processing sales, Crown's growing, global network of equipment manufacturers moved fabrication closer to its overseas markets, too.

Anticipating unification of the European market in 1992, George Anderson told the *Star Tribune* that Crown was open to new applications of its extraction engineering expertise overseas. Among the possibilities, Anderson cited new ways to extract pollutants from soil or industrial byproducts and applications in the chemical and pharmaceutical industries. "A large part of what we do," George Anderson said, "is customer-driven."

On The Road. Again.

Only six years after moving into new headquarters at 1600 Broadway in Northeast Minneapolis, Crown's growth called for yet another relocation in late 1991 to 2500 West County Road C in Roseville, a suburb of Saint Paul. In that new space, Crown had 23,000 square feet of office and warehouse space to house Crown Holdings, Inc., Crown Iron Works and the new Wurster & Sanger division. Always careful with finances, the leaders of Crown made sure that renters in both locations, new and old, generated income to help Crown cover its operating costs. "Our new building is a handsome one," Cliff Anderson said, "and with extensive (expensive) remodeling, it will be a headquarters to be proud of once again. With a little luck to go with our efforts, maybe we can some day fill our new building, too."

Paul Ell remembered collecting rent checks from the Tyler Street plant before Crown sold that property in 1986. "I joined Crown in 1980 and when we closed the structural steel operation at Crown, we created artists' lofts, the first in Northeast Minneapolis, over one of the bays in the fabrication shop at Tyler Street," Ell says. "To generate income from the empty space, we rented these lofts to young artists just getting started and one of my jobs was to collect rent each month." Ell says he was always impressed with Crown's frugal nature, an attribute that served the company well during economic turbulence of the 1980s and 1990s.

"We're very good. Let's keep getting better!"
Cliff Anderson in the Spring 1991 issue of *Crown Gazette*

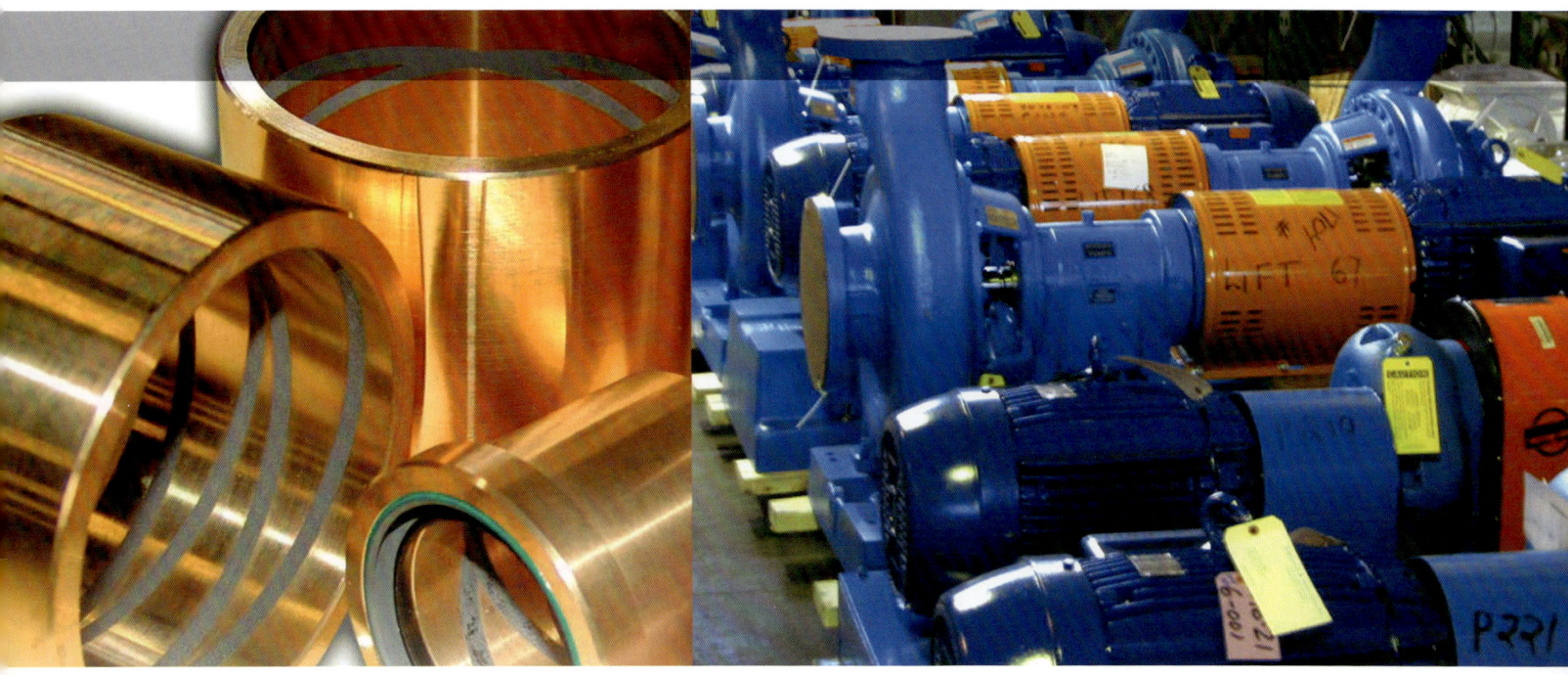

(Above, Left to Right): Crown's modest Spare Parts Department grew so much that it became a division. These are only a few examples of thousands of parts in the Crown inventory: bronze bearings for a DT (desolventizer-toaster), centrifugal pumps for a solvent extraction plant, rollers for the Crown Jet Dryer and part of a 12-foot-diameter sprocket that drives Crown's 6,000 tons-per-day extractor.

Initiative Pays

What made Paul Ell's career at Crown was not financial rigor alone, however, but the chance to become an "intrapreneur."

Ell had traded his credit manager hat in Crown's export business for management of Crown's Spare Parts Department, a modest little operation that supplied goods as small as a six-inch diameter "sight glass" for a liquid container to a 20-foot-high carbon steel section for an oilseed extractor. Every spare part was made to precise measurements called for by Crown's engineers and customers. To supply spare parts, Ell called on about 25 primary suppliers and another 100 specializing in ancillary equipment.

"Cliff Anderson came to me in 1990 and said, 'I think we can make some money on this spare parts business. Why don't you pursue it, Paul, and see what you can do?' Cliff's confidence gave me a good feeling and I was excited about the chance," Ell remembers. At the time, Crown's parts business was generating about $880,000 in annual sales.

As Crown's customer list grew in numbers and project scope, there were more requests for spare parts, replacement parts and maintenance parts. In addition, as Crown's projects became more global, Ell's department helped supply new equipment for some plants.

Ell's parts vocabulary expanded fast. "I was trained in finance and I had to learn engineering terms," he says. "When I first started, I walked into a plant and asked, 'Where's the extractor?' I was standing right under it." Ell and his colleagues, Linda Moseley and Chris Johnson, developed a "cheat sheet" to keep track of parts names and their uses. "Depending upon the type of extraction they were doing, an engineer or project manager would call a part by a different name," Ell says. "One guy might call some piece of equipment a 'first stage,' another called the

same thing an 'evaporator' and the next guy called it his 'first effect.' I spent a lot of time in our engineering department learning terminology."

Paul Ell and his colleagues began developing good rapport with Crown's customers and suppliers. As calls came in, engineers and project managers referred them to Crown's ambitious little parts department. "We started quoting prices and selling parts and working hard to find answers to our customers' questions," says Ell. "It mushroomed from there."

By 1993, Ell's department had multiplied its revenues and was renamed the Renewal Parts Division. In 2007, as Paul Ell looked back on his chance at intrapreneurship, he pegged his division's annual sales at about $4.2 million. "People never get stale at Crown," he says. "There is always a new challenge, something to look forward to. I was encouraged, but I was left to do what I thought was best." Still general manager of his division, Ell attributes Crown's unusually low turnover rate to this supportive and trusting management style. "Once you become a Crown person, you stay a Crown person," he says, smiling. "Either you're gone in two years or, like me, you're here for decades."

Good Faith Rewards

Richard Ozer joined Crown with similar aspirations. "Crown wanted to expand into nontraditional areas of specialty extraction using its engineering expertise," says Ozer, a trained engineer who worked in specialty chemicals for 20 years before joining Crown. "I was impressed that Crown was an innovative company willing to look at different applications that others might not consider. Crown's investment in specialty extraction has been a long-term gesture of good faith."

> "The first two Crown Model IV extractors for the pharmaceutical industry were installed and started up."
>
> ***Crown Gazette,*** **Christmas 1991 issue**

This Model IV Extractor was sold in 1989 for a specialty extraction plant for Evening Primrose, the first nutraceutical run in a Model IV.

> "Specialty Extraction can be anything from food-related extraction, to fragrances, to pharmaceuticals, to environmental cleanup—the customer names it, R&D works on it."
>
> ***Crown Gazette,*** **Winter 1998 issue**

The first, big payoff came in the 1990s when Crown produced soy protein concentrate (SPC) for human foods such as processed meats and veggie burgers, and high protein feed for animal consumption, including soy protein for aqua farming.

Highly effective as a nutritional source, soy protein concentrate provides about 90 percent of the proteins required by the human body. It is being used more and more to enhance meat flavor and moisture content, while serving as a "product extender."

Crown had been introduced to SPC in 1970 when George Anderson adapted an early Crown extractor for the first significant SPC commercial plant in Israel, developed by Daniel Chajuss. "Daniel invented much of the SPC technology," says Anderson. "He and his father became the pioneers in SPC production worldwide, and we have collaborated with Daniel since that time."

The invention of this extractor led to a succession of plants that Crown built for ADM and others through the 1980s. By 2007 nearly 80 percent of SPC was made with Crown equipment.

"The demand for high-protein food is increasing worldwide," says Ozer, "and fish is a less expensive protein to produce than beef and poultry. Between 2000 and 2010, the aqua culture market is expected to multiply three times in total tons produced per year."

From Bulbs to Beer

Crown ventured into all kinds of specialty extraction through the 1990s. They investigated ways to reuse rendering fats for animal feed and soaps. They applied their know-how to help Xerox Corporation develop a better process for making copying toner. They used their drying techniques to produce fine, white film that coats the inside of light bulbs. They explored converting waste material from steel mills to a tough pigment used to protect industrial floors and road beds. They helped Budweiser make a silica gel filtering agent for their beers. They worked with a nutra-ceutical company to extract a component of Evening Primrose that helps treat psoriasis and they teamed with a baby food manufacturer to produce synthetic Omega 3 oils for their products.

Crown's expertise in extraction, drying and desolventizing led to these new applications. "We were able to demonstrate our versatility," says Ozer, specialty product sales manager, "and with a separate department focused on specialty uses, we've been discovering new applications."

Crown's Global Voices

Bolivia, Barbados, China, Arkansas, North Dakota. Crown people traveled to destinations on many continents. The voices of Crown people became more diverse and multilingual. Passport stamps multiplied for sales people, engineers and project managers.

When planning domestic travel, Sharon Trocke learned to ask the destination state, as well as the city, when making reservations. "There are a lot of Decaturs in the U.S.," she says. "When we started with more international jobs, it was a challenge to understand visa requirements, airport locations and train schedules to all the agricultural areas."

Recognizing the terrain of Crown's "global village," Cliff Anderson suggested total immer-sion Spanish courses for some U.S. employees, while others packed foreign language dictionar-ies and learned on the fly.

These overseas assignments became the stuff of vivid memories. Jeff Christopherson, who joined Crown in 1979 as an assistant project manager, remembers his first plant start-up in northwest Beijing, China, in 1991. "I was there with John Chasteen, a field technician who hailed from Kentucky, and I was a draftsman from South Minneapolis. I wondered how the two of us could be sitting on a train chugging across China after just starting up a huge plant. It was a big responsibility. The company had faith in us and that was awe-inspiring."

Christopherson and Chasteen happened to visit China during the Chinese New Year. The entire plant staff (and country) took the day off to celebrate. The American pair was invited to share New Year's lunch and they learned to make Chinese dumplings. In fact, a local TV crew filmed their cooking lesson, along with Christopherson and Chasteen serenading their hosts with "Auld Lang Syne."

Christopherson also recalls a 26-hour train ride across Russia and a sauna in Rostov-on-Don, near the Black Sea, for a business meeting. "We started with a sauna and it was culture shock for me," he recalls, "sitting with a group of naked Russians I'd never met."

From Indonesia to Mexico

Project sites are often distant and isolated. One such location was a bitumen mining site on an island off Indonesia. This natural resource was the key ingredient in asphalt and Indonesia's bitumen was especially desirable because of its durability. To reach that site in 1998, Crown's Dr. Bob Wills, R&D manager, and consultant Larry Hastings traveled two days. They endured four airline flights, a five-hour boat trip and nearly two hours in a Land Rover.

"I went all the way around the world in five days with Barry Smith, our Wurster & Sanger division general manager. Now I know what jet lag is all about!"
Cliff Anderson, *Crown Gazette*, Summer 1992 issue

"When I started with Crown, we were 'third fiddle' in the extraction business. Now we're number one. We've outlasted everyone. . . . because we're better."
Jeff Christopherson, design manager, extraction

(Above, Left to Right): Clifford Anderson in China meeting with leaders of CNTIC. Jeff Scott on a sales trip selling Crown's first job in mainland China, a cottonseed operation in Jinan.

> "Jabon la Corona opened the market for Crown in Mexico."
>
> **Haskell Cooke, business development manager, Mexico**

While Indonesia beckoned, Haskell Cooke, a biochemistry engineer with extensive experience in edible oils and specialty food products, scouted new business opportunities in Mexico. Cooke was based in Hermosillo in the Mexican state of Sonora, an area known for its soybean, cotton and safflower crops. "R&D Equipment Company worked with Crown in Mexico and they asked me to represent Crown there," says Cooke. Travel and cultural differences weren't the issue for Cooke, but name recognition was.

Crown Iron Works was an unknown in Mexico and two competitors, DeSmet and French Oil Mill, dominated. That changed when Jabon la Corona—a huge, century-old oleochemical producer—hired Crown to design a new canola processing plant. "The owner of la Corona asked Ralph Romero, from R&D Equipment, which company had the best technology in solvent extraction worldwide," Cooke recalls. "He said 'Crown.' We promised on-time installation, the best quality and maximum production in just one week after start-up. The la Corona plant became our reference plant in Mexico. It was open for everyone to see."

Doing Business in Russia's "Wild West"

When Crown created Europa Crown Limited with a group of former Rosedowns engineers, the decision was courageous, says Richard Young, one of those engineers. "Small, family-owned American companies can be insular. They're not inclined to think globally," says Young, Europa Crown's engineering director in the United Kingdom. "Crown wasn't like that. They worked with us, instead of setting up their own, separate company. They wanted to sell and fabricate an entire product line, not just the extractor. When Europa Crown started, we had nothing. Crown's decision was brave."

(Above, Left to Right) Valentin Zaletkin, Manager of Crown's office in Moscow. Boris Solovyov, an engineer at the Moscow office. A processing plant in Central America.

Europa Crown opened up new areas of the world, including Russia, not long after the demise of the U.S.S.R. Lloyd Walker, one of Europa Crown's original employees, remembers Crown's first extraction plant contract in Rostov-on-Don in southern Russia. "We were to be paid for our work by an agent in Austria," he says. "When it came time to ship our equipment, we were expecting a big payment in dollars, but our Austrian contact didn't have the money." Crown's fee would not be forthcoming, the Austrian explained, until a big shipment of wheat arrived in China and was sold. "'Once that ship gets to China, I get paid for my wheat,' he told us, 'and then I can pay you for your extraction plant,'" Walker, Europa Crown's commercial manager, remembers. "It was a worrisome time, to be honest, but it did work out."

Europa Crown kept a Moscow office open through the "wild west" years of Russia's evolution from a Communist state to a free market economy. By 2000, Russia's demand for oils and fats had dramatically increased as the daily diet for Russians improved and Crown was the happy beneficiary of that lifestyle change.

Birthday Greetings?

Crown celebrated its 120th anniversary in 1998 and with it came a dose of market reality. "What a difference a year makes! I'm sure by now that everybody knows our work load is considerably less than it was this time a year ago," Jeff Scott, leader of Crown Iron Works' largest division by far, commented in the *Crown Gazette*. "This slowdown is mainly due to profit margins our customers make on each bushel of grain they process. A year ago, margins were over $1 per bushel and today they are below 25 cents! Most of our customers are losing money or breaking even, at best. Most companies will not consider installing new plants or spending capital to upgrade existing facilities."

Business à la Barbecue

Like Crown, Intecnial was a family business. Augusto Ovidio Skrzypek had just received his degree in electrical engineering in 1971. Augusto's two brothers and his cousin—also with technical backgrounds—saw opportunity as industries in southern Brazil expanded. "We had many meat processors in our region," he says, "and other companies working in the food industry. They needed equipment for electrical panels, cable installation, tanks and vessels and other machinery."

When potential customers asked the young entrepreneurs if they were up to the job, the immediate answer was "yes." "We built our own workshop and started manufacturing, calling ourselves Intecnial," says Skrzypek, "and we started to diversify, from electrical to mechanical contracts." The Belgian behemoth DeSmet came calling in the 1970s. "Can you produce machinery?" DeSmet asked. The eager engineers again answered in the affirmative.

"We started learning a little bit about extraction and the soybean business in Brazil," says Skrzypek. DeSmet hired Intecnial to handle site installations and Skrzypek and his partners gained expertise in cables, wires, piping, machinery and plant structures.

Then Skrzypek met Jeff Scott of Crown Iron Works in the early 1990s. "Jeff was in Bolivia to sell extraction plants to a local company," he says, "and we were working on the electrical and site installation." Before that meeting ended, Skrzypek asked Scott to visit Intecnial in southern Brazil.

A few years later, when Coopersul, a successful Brazilian co-operative, investigated suppliers to build a new soybean extraction plant, they resisted hiring DeSmet. "They told me they weren't happy with the DeSmet process," says Skrzypek. "They decided to hire Crown as long as the equipment was manufactured in Brazil.

"Jeff Scott sent his engineers to Brazil to qualify three possible workshops to partner with them," he says, "then Jeff flew to São Paulo, the largest city in Brazil, to visit the workshops and make a decision." Scott was on his own. Crown's local consultant, Ricardo de Oliveira Carvalho, who supplied the fabricators' names—including Intecnial—had died within days in a car accident on a winding, Brazilian road.

"I got a map out," says Scott. "One company was in São Paulo, another was in central Brazil and Intecnial was way down at the bottom of the country. At least one thousand miles away. I just crossed them off my list."

Word of that decision traveled fast through the grapevine of Brazilian business and the boys at Intecnial snapped into action. They dispatched their colleague Werner Wagner to find Scott and urge him to make the long trip. Wagner took Scott to dinner and made his plea: "Werner said, 'You've got to meet our company. And besides, they're going to kill me if I don't get you to come down,'" Scott recalls.

Wagner had two tickets to fly the next day.

It was Sunday. When Jeff Scott arrived, he immediately remembered Augusto Skrzypek from their cordial meeting in Bolivia. Augusto, the consummate host, had a special Brazilian barbecue dinner prepared and Scott was the guest of honor. In short order, the Yankee learned the extensive capabilities of Intecnial and the partners who owned it. "Jeff will tell you that I kidnapped him," Augusto grins. But the outcome was a win-win proposition. They worked together on the 1,500 tons-per-day Coopersul soybean processing plant installed in 1993.

Intecnial has represented Crown Iron Works in Brazil, Bolivia and Argentina from that time forward, selling and manufacturing Crown equipment. There were years when work with Crown generated 50 percent of Intecnial's revenues.

"Brazilian business is very much based on relationships," says Paulo Telles, a native of Brazil and a Crown engineer. "It is not only about pure business, but it is social and personal. That explains why Augusto will fly to Minnesota in the dead of winter to have face-to-face contact with Crown people."

"Our two companies are going in the same direction in the same business," says Skrzypek. "We have trust because we are both family-owned and we have taken the time to build a strong relationship."

Crown and Intecnial of Brazil forged a working relationship in the early 1980s that continues today. (Left to Right): Marcio Teijeira, George Anderson of Crown, Romualdo Skrzypek of Intecnial, Hector Autino of Bunge, Allen Ost of Crown, Osvaldo Pioli of Bunge, Alcir Dall'Agnol of Intecnial, and Hernán Paredes of Crown.

In his market lesson for Crown employees, Scott explained that processors tend to build plants on boom or bust schedules with rapid deliveries and tight shutdowns, even in the best of times. "Profit margins increase and decrease based on the supply and demand of oilseed products worldwide," Scott said. Because many new plants were installed from 1993 to 1998, Scott said, supplies of meal and oil exceeded demand, and profit margins decreased.

Crown was not alone. "Our competitors are in the same situation," Scott said, "and because there are fewer projects to pursue, our competitors try to attract customers by reducing their selling prices."

But Scott was optimistic: "Long term, there are many solid reasons to expect a long steady increase in the average consumption of vegetable oils," he said. The slowdown at Crown would just give the company time to get even better, he reasoned: "If anyone sees a way to improve a product, speed up the response to a customer, improve our quotes or literature, anything you see that might make us more able to serve our customers, tell us! If you have a useful idea that would improve something we do, why not volunteer to do it?"

Cliff Anderson was equally positive about his company's future: "Crown has the finest products, the most expertise, and the best reputation in the oilseed industry," he told employees in the spring of 1998. "This industry has a certainty of long-term growth. And if you are a contributing part of an organization that can claim a piece of that action, you should have a very good future."

From Detergent to Diesel Fuel

Hernán Paredes of San Pedro Sula, Honduras, took those appeals for initiative to heart. Having worked with Crown's Wurster & Sanger division, Paredes, a chemical engineer, had extensive experience with oleochemicals and glycerin production. When he joined Crown in 1995, he was assigned to promote Crown's expertise in Latin America, overseeing new projects from sales to engineering to project management. "Crown didn't have much business in Latin America," he says, "but we had all the technical capabilities."

After Crown won a contract to do preliminary engineering work for a methyl ester glycerin plant in Comayagua, Honduras, Paredes was the project's "midwife." It was a first for Crown and the large, complex project held promise. The customer was Cressida, a major, local company.

"Methyl ester is the chemical name for biodiesel," says Paredes, "but in those days, it was used to produce detergent. I thought that methyl ester had a good future because it could replace petroleum in detergent manufacturing." Producing biodiesel fuel was almost an afterthought.

"Biodiesel was being produced in Europe, but it wasn't in America in the mid-1990s," says Paredes, Crown's Latin America business manager. "Nor was it produced in Central and South America. When we built the plant for Cressida, it was apparent that we had another application: producing biodiesel. It was the same process."

The Cressida project also included a palm oil refinery that would incorporate Crown's new Diflow deodorizing technology for edible oils, a process developed by the engineers of Europa Crown Limited.

Hernán Paredes joined Crown in 1995 to promote the company's expertise in Latin America.

"If this project goes through, it means a $6 million-plus contract for us . . . and a key ingredient for a new detergent formulation that could revolutionize the way we wash our clothes."
Crown Gazette, **Spring 1998 issue**

Shown here are decanters used in Crown's biodiesel process to separate glycerin from biodiesel.

"This installation is big news in the detergent industry. We have big hopes that the Cressida project will open more doors for Crown in the near future."

Crown Gazette, **Winter 1998 issue**

Crown and Hernán Paredes were riding a wave of optimism with high hopes for the Cressida plant. Then the bottom fell out. Giant Unilever bought Cressida, closed the plant and the equipment was dispersed to points unknown. "They shipped off everything—some equipment to Guatemala, some to Costa Rica, some to Ecuador," Paredes remembers. The plant never started up but there was no reason it wouldn't have worked. Crown had its chance again in 2000 with a detergent manufacturer in Houston, Texas.

That opportunity would soon propel Crown into a brand new industry in the new century: biodiesel production. In its infancy in America, industry experts predicted that only two million gallons of biodiesel would be produced in 2000. But by 2007, that projection would skyrocket to 150 million gallons. And Crown would be there—front and center.

A Genuine Win-Win Equation

By the decade's end, two newcomers joined Crown in senior leadership positions: Gary Koerbitz, filling a new position as vice president of operations, and Ralph Romano, Jr., vice president of finance. Seasoned professionals, they brought industry and financial management experience to Crown.

Their impressions of the company offer fresh insights into its culture.

Koerbitz, a chemical engineer from International Falls, Minnesota, knew Crown as a customer—first as a production assistant with ADM and later as an associate process engineer rising to vice president of manufacturing for Harvest States Cooperative (Honeymead). "At first, all I knew about Crown was its nameplate on a piece of their equipment," says Koerbitz. "Later, I was evaluating their equipment and purchasing it." He remembers Crown's first hot dehulling plant in Indiana and Crown's "beta site" for its first, full-blown hot dehulling plant at Honeymead in Mankato, Minnesota. "We were fully confident that it was going to work," Koerbitz recalls, "but there was still that risk of trying something totally new."

"Urning" Confidence

Fifty-four years after he bought a Crown product, George Emerson was disappointed. And he said so. The product was a No. 2A Eterno Crown cast bronze urn and it graced his mother's grave from 1941 until 1995. When he visited, Emerson discovered one of the urn's handles on the ground.

Emerson had saved the original sales literature and he read aloud to Crown's Sue Young: "Try now to imagine a hundred years hence. Then future generations will pause to admire the tokens of memorials built today. Other materials may have crumbled but bronze will rigidly survive. . . . " Well, Emerson mused, that was not quite the case.

Even though Crown had exited the ornamental bronze business 35 years earlier, Cliff Anderson asked ever-resourceful Bill Kratochwill to investigate. He soon discovered that the screws holding the handle to the urn had rusted away. Kratochwill took the handle back to Crown, retapped the holes and substituted stainless steel screws. Mission accomplished. (Colleagues ready with a joke assured Kratochwill that he had finally "urned" his keep.)

Product reliability and conscientious follow-through are Crown tenets. "Our stuff works," Cliff Anderson says simply, "and if it doesn't work, we make it work or replace it. We have a reputation for that. In this field of processing equipment, customers are facing large dollar investments. They want to know that when they spend millions on a process, it's going to work."

Hernán Paredes, Crown's Latin America business manager, echoes this commitment: "We are a very serious company and our customers know their investment is safe with us. Our products and service are first quality."

ETERNO - CROWN

Distinctive Metals of Permanent Beauty

Richard Young represents Crown's engineering for Europa Crown Limited. "I firmly believe that Crown's extraction plant is the best in the world," he says. "It has so many merits that no one else can match. We stand on reliability and performance. Our process is elegantly simple . . . and it will work forever."

But reliability and performance are not limited to design and equipment. They characterize service, too.

Hernán Paredes has served as a project designer, manager and plant start-up supervisor many times. Preparing for the start-up is like counting down to the lift-off at the Kennedy Space Center. "It's not an eight-hour job," he says. "Some days are more than 16 hours. Depending upon the size of the plant, the time could stretch from three days to a month. We just go to a hotel, sleep a little bit and return. We have to be prepared because there are so many things to consider. Everything is not perfect. We have to be resourceful and think fast. It's fun. Tiring, but fun."

Alcir Dall'Agnol, an engineering supervisor for Intecnial, S.A., has witnessed many plant start-ups for Crown. One of his favorites involved spontaneous invention. "It was Saturday night and we had a problem with one machine," he says. "I made contact with someone from Crown and we talked the problem through. We solved it. Then I asked one of the engineers from Intecnial to help me design the new equipment part." That was Saturday.

Early Sunday morning, Dall'Agnol called an Intecnial factory worker in Erechim and explained what he needed. He sent the equipment part by car, a six-hour drive. In one day, the part was re-made and rushed back. Dall'Agnol and some of his colleagues didn't sleep for 48 hours. "We had the plant up and running by late Sunday night," he says. "Because of our solution, the customer ordered another plant. Service is 70 percent of all new sales."

Koerbitz is convinced that Crown got the job with Honeymead because of relationships forged 20 years earlier, in the 1970s. "It went back to George Anderson, Jeff Scott and Glenn Brueske of Crown and Jim Amlie and Mike Cheney of Honeymead," he says, "and their long-term relationships of trust. That's what a lot of business is about. Once the Mankato hot dehulling plant was up and running, that's when ADM, Cargill and everybody else got interested." Koerbitz witnessed the collaboration of Crown people with Honeymead staff. "They had a goal and they worked together to accomplish it. When you invest millions and millions of dollars, nothing means more to management than everything working."

Next, Koerbitz oversaw negotiation, planning and construction of a 4,000 tons-per-day Crown hot dehulling and extraction plant in Fairmont, Minnesota, for Harvest States.

When the project was put on hold—a decision Koerbitz opposed—Crown's Jeff Scott queried him about a new position at Crown. The company was growing at a fast clip and they needed a vice president of operations. Koerbitz joined Crown in 2000 and he was surprised most by its international scope:

"I assumed a majority of Crown's business was in the U.S.," he says, "but I discovered that more than 50 percent was international. Crown went from being an iron works, to an extraction equipment company operating mainly in the U.S., to a global company with multiple divisions. They risked going to China and South America. They went out on a limb to make it work. Cliff and George supported that growth and Crown's people did the ground work and made it happen." Vital to this growth, Koerbitz says, were the partnerships that Crown forged around the world and its dedicated employees.

Ralph Romano, Jr., a native of Duluth, Minnesota, trained in accounting at the University of Minnesota Duluth. Before joining Crown in 1999, Romano worked with five Minnesota companies, including Jeno's Pizza, Polaris Industries and Bay West, an environmental services firm.

Romano recalls Cliff and George Anderson's kindness as Romano's predecessor, Bruce Wandrei, struggled with cancer. "I quickly saw that they were genuinely caring people. I also discovered that making money was not their main focus; it was instead treating employees and customers with equal respect."

In Jeff Scott, Romano saw a visionary concentrating on building strong relationships. "Jeff created partnerships built on trust, not ownership," says Romano. "As a result, Crown grew without investing heavily in opening offices around the world. Historically, most companies that went international were large companies with people and resources. We didn't have either, so Jeff found the best local people that he could trust. That approach made us nimble; when times were tough, we didn't have a lot of extra costs to cover."

Partnerships and business collaborations work for Crown, Romano says, "because Crown is not a control culture. People have a lot of freedom to do what they need to do." The partnerships are mutually beneficial arrangements. "Our partners understand that they're making nice money," says Romano. "I've worked with people who say win-win, but what they really mean is, 'I win, you lose.'

"With Crown, it's a real win-win."

Feeding and Fueling the World

It was a golden opportunity for Crown. ♛ West Central Cooperative, a 68-year-old company based in western Iowa, was a pioneer in the American biodiesel industry. They had been converting soybean oil to methyl ester since 1996 and—with their trade name product SoyPOWER™—they were leading an embryonic business. Even so, they knew they could be even better. ♛ By 2001, West Central wanted to build a larger and better biodiesel plant and they asked Crown to do some design work.

A field of canola plants. (Top): The interior section of a Crown biodiesel plant, developed with computer-aided design. (Bottom): Visiting a Crown Friendship Engineering Plant in Wuhan, China. (Lower Right): Photo image of Crown biodiesel equipment.

> "Crown had a very good reputation in soybean processing and extraction, but didn't have much experience in biodiesel. We learned the business together and we grew as we learned more and more."
>
> **Jeff Stroburg, CEO and chairman of Renewable Energy Group**

> "I worked with Cliff Anderson to craft the relationship with West Central. I discovered that if he tells you something, you can take it to the bank."
>
> **Jeff Stroburg**

"Crown was known for its expertise in process engineering and equipment," says Myron Danzer, vice president, customer and technical services for Renewable Energy Group, Inc. (West Central Cooperative is their parent company). "Crown designed the process we needed to meet the biodiesel specifications we aimed for," Danzer says. "Our goal was to make sure our product met the quality and cost requirements of this new industry."

Not long after Crown delivered these recommendations for improving and expanding West Central's biodiesel process, an explosion crippled their original small plant.

"They asked us, 'If you had to start over, rather than re-engineer what we already have, what would you do?'" says Derek Masterson, Crown's product sales manager—biodiesel.

"We essentially took our methyl ester production technology and adapted it to their needs," Jeff Scott told *Biodiesel Magazine*. "The process we offered was modified based on a joint effort between our company and theirs."

Crown's process had distinct advantages. Crown designed a fully continuous operation—not a batch process—to produce high yields in each phase of production. Based on performance, it rivaled the best in the United States and Europe. In addition, Crown's design reduced emission losses and efficiently recovered methanol and water used in the process. It also incorporated a glycerin treatment step to produce a standard, crude glycerin. Because the process involved flammable solvents such as methanol, Crown built in strong safety features.

"Our experience reduces risk . . . that's another thing that sets us apart," Jeff Scott told *Biodiesel Magazine*. "Crown's biodiesel plants are designed for maximum efficiency and safety."

West Central's new plant in Ralston, Iowa, opened in the winter of 2002, producing 12 million gallons a year. "It is one of 18 biodiesel plants nationwide and the largest in the U.S. dedicated to biodiesel production," the *Crown Gazette* reported. "More than a dozen others are planned by various private companies and farmer cooperatives. The plant also represents one of the first projects Crown has done entirely in 3D."

Crown's Bill McDonald and Jeff Kraker used several CAD (Computer-Aided Design) applications to create the plant schematic, exact to the tiniest of details. They even "installed" ceiling lights that cast shadows in the 3D version and pictured two inspectors who looked a whole lot like McDonald and Kraker peering at equipment on the plant floor.

Gazillion Gallons

With a successful, new plant on their hands, West Central Cooperative and Todd & Sargent, the builder, formed their joint venture called Renewable Energy Group (REG) in 2002, with Crown as REG's sole process technology provider. This trio offered everything from biodiesel plant site selection, design and construction to plant management, procurement and marketing. (By 2007, REG had more financial partners, including huge Bunge Oilseed Processing, and 16 plants in some stage of planning, construction or start-up all over the United States.)

In fact, it wasn't long after West Central's Ralston plant opened and REG arranged for Crown to build a 30-million-gallon plant for SoyMor in Albert Lea, Minnesota, that Crown was tapped to build another plant for Minnesota Soybean Processors in Brewster. Crown actively sought contracts for 10- to 30-million-gallon plants with aggressive timelines: "For a greenfield

Learning From Europe

Biodiesel came mostly from Europe. Rudolph Diesel, a German inventor and engineer, designed the compression engine in the late 1880s and first used peanut oil to fuel his invention, followed by vegetable oils. With the rapid commercialization and reduced price of petroleum oil in the 1920s, however, the use of vegetable oils in diesel engines declined.

During World War II and again during a petroleum oil shortage in the 1970s, more modern formulations of biodiesel had some commercial success—largely in Europe where biodiesel was recognized as a smart and environmentally friendly fuel choice.

> "West Central had the foresight and ambition to mimic some of the players in America's ethanol industry."

Greg Waranica, Crown sales manager—oils and fats

"Germany and Austria have had a thriving biodiesel industry for more than 15 years," says Greg Waranica, Crown's sales manager—oils and fats. "Dr. Joosten Connemann, a German business owner, was competing with the big companies like ADM and Cargill in oilseed processing and he wasn't faring well. He saw an opportunity in biodiesel because half the cars in Germany were diesel. You could call him the 'godfather' of biodiesel in Europe. Germany subsidized the industry and producers were largely small farmer groups."

While Europe embraced biodiesel, America focused on ethanol production. Americans more familiar with ethanol know that it is a product of fermentation and distillation, producing alcohol. Corn is the primary crop used in making ethanol, but any material rich in starch is a likely candidate.

When the 1990s dawned, the modern biodiesel trend had a tremendous boost all over the world—including America—when China and other emerging economies began competing for fuels and oil prices spiked after the events of 9/11.

In 2002, Minnesota's legislature was the first to pass a law requiring that diesel fuel contain at least 2 percent biodiesel. Since that time, dozens of other states, including California, pursued similar legislation. By 2008, four other states had passed similar mandates.

When Congress passed the Renewable Fuel Standard to stimulate biodiesel production, that was the jump-start, says Waranica. "The legislation was effective in 2004, orders were placed that year and plants were built and started-up beginning in August 2005. That's when the biodiesel craze took off worldwide. South America and Southeast Asia joined the race in 2005 and 2006."

The first American company to produce biodiesel commercially was Kansas City, Kansas-based Interchem Industries in 1991. Their product, called SoyDiesel, was made on a small scale to supply a handful of biodiesel research projects and a few small cooperatives in the area. "They had small diesel trucks that they drove all over the U.S.A. advertising the new product," says William Shurtleff, founder of the Soy Information Center, a massive database and research consultancy devoted to the industry. "In 1991, there was a glut of soybean oil on the market and it was depressing the price of soybeans," he says. "The soybean associations looked for the next, best way to sell their crops. Bill Ayres, an owner of Interchem, became the key supplier and Kenlon Johannes, of the Missouri Soy Association, formed what became the predecessor to the National Biodiesel Board. They were the spark plugs. After that, state associations began contributing to early biodiesel engineering research and much of that early work was done by Crown Iron Works in Minnesota."

> "When you have clients that are already crushing oilseeds for the same type of oil being used for biodiesel production, it makes sense to take that extra step."

Jeff Scott, quoted in *Biodiesel Magazine*

Crowning
Achievements

Minneapolis-based Crown Iron Works is dedicated to continually improving its biodiesel process tech...

By Tom Bryan

It would be perfectly logical to assume that Crown Iron Works builds biodiesel plants.

The Twin Cities-based company has a rich history in iron casting and steel fabrication. It provided some of the structural steel for the tallest building in Minneapolis, the IDS Tower, along with the city's downtown walkways and the Metrodome, home of the Minnesota Twins and the Minnesota Vikings. Couple this portfolio of work with 60-plus

Crown Iron Works is today a full-service engineering firm that provides process technology and equipment designs for oil and oilseed processing. Biodiesel process technology and production equipment is just one facet of the company's burgeoning Oleochemical Division, and an area Crown has heavily invested itself in.

In the biodiesel industry, Crown is probably best known for providing the process design and equipment for one of the nation's

under construction 30-m... Soybean Processors plant in proposed 30-mmgy SoyMe... Lea.

So how does an oilse... refining equipment supplier f... production in just a few ye... ple answer is synergies-en... commodities between two... es-and working with clients t... oil and oilseed crushing facil...

These images are not photographs, but computer-generated interiors of Crown biodiesel plants (complete with technicians on-the-job), made before they were built. 3D computer-aided design, led by Jeff Kraker, is commonplace at Crown today.

> "One of our real growth areas for biodiesel is Southeast Asia. It's the biggest palm growing area in the world and palm oil is an excellent feed stock for biodiesel."
>
> **Dan Anderson, Crown's director of Asian operations**

30-million-gallon plant, it would take 12 to 14 months," Jeff Scott told *Biodiesel Magazine,* "It takes five to six months to deliver the equipment. For an existing oilseed processing plant that already has some of the infrastructure, the time frame could be faster."

Within a few years, Crown was exploring alliances with several other companies in fuel and bio-chemical industries related to biodiesel.

Study a map of biodiesel plants operating and projected from year to year; the numbers are stunning. *Biodiesel Magazine* reported U.S. plants producing about 250 million gallons in fiscal year 2006 (ending September 30) with about 77 million gallons of those coming from Crown plants. Total output in the United States doubled each year in the previous two, 2004 and 2005.

Did that mean that production could reach 350 million gallons in 2007? Crown's Derek Masterson answered that inevitable question: "There's enough capacity to make that possible and half of it could be from our plants." By 2008, Crown had sold 21 plants with six outside the United States—one in Malaysia and five in South America. "Six is a small number compared to the total number of biodiesel plants operating internationally," he says, "but we started at zero." Crown's international contracts were a direct result of its domestic success. The company faced stiff competition worldwide from Lurgi, DeSmet and Westfalia—all large and well-established in the industry.

Wired and Willing

Crown's long-held relationships around the world make its entry into the international biodiesel business possible. Crown Project Engineers Bill McDonald, Kathy Liesmaki and Ryan Popinga are working on a turnkey biodiesel and glycerin refining plant in Malaysia for Mission Biotechnologies. Pisces Engineering in Malaysia is handling the equipment manufacturing

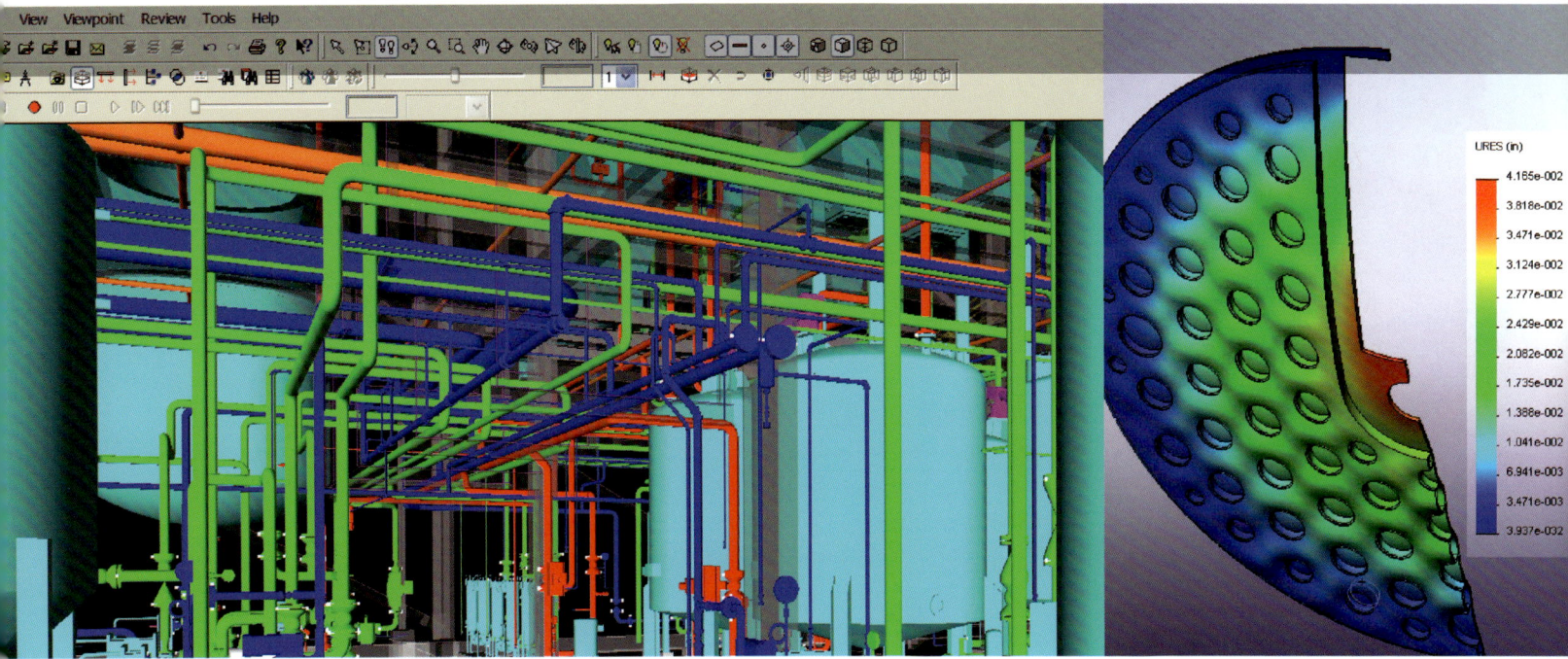

and installation, while Crown is doing the process engineering and oversight. "Mission Biotechnologies is responsible for the building footings," says Popinga, "and we're responsible for everything from structural steel design, to erection, installation of process piping and vessels, electrical installation, and process controls—everything involved with the process buildings is in Crown's scope."

And they're doing virtually everything from Roseville, Minnesota. "Before the Internet, we would have had someone physically on-site in Malaysia at all times," says Popinga, "and that's certainly not the case now. Our technology allows Crown to cast a wider net and win business in places where it would have been much harder to conduct business before."

In December 2006, Crown was overseeing installation of the plant's structural steel and delivery of the first process vessels to the site. By that time, Popinga had 2,513 e-mail messages in his correspondence folder, some multiple pages long and packed with specifications and mind-boggling detail. "We're talking about thousands of questions and answers," he says, "from our fabricator and from our customer. Thanks to the technology we have, I'm receiving photos from Pisces and I can respond almost immediately via e-mail."

Crown people have remote computer access to all project sites, regardless of their physical location. "We can be 'in the office,' virtually anywhere," says Popinga.

All Points East

By the first decade of the 21st century, Crown's international business represented as much as 80 percent of the company's annual revenues. "The U.S. oilseed industry used to be a domestic and export market where they sold the raw materials and finished products overseas," Jeff Scott

(Above): Crown was one of the first in its industry to use Finite Element Analysis, an advanced design technique that can pinpoint how and when a machine part might fail and how it can be strengthened. Crown employee SueEllen Altrichter provides this expertise at Crown.

> "It looks as though we should be busy in China for the next few years."
>
> **Jeff Scott, in spring of 2000**

said in 2000. "Now countries in the Middle East and Asia are building their own plants and processing locally grown or imported products. Our largest processors in the U.S. adjusted to the changing times by building or buying plants overseas, too."

Crown was ready to capitalize on this shift with strong international visibility through Europa Crown Limited and Intecnial in South America, matched with equipment fabricators with strong track records in Malaysia, Brazil and China.

Crown landed its first complete preparation and extraction plant with East Ocean Oils of China in spring of 2000, followed only a few months later with confirmation of two more. "In addition to East Ocean Oils, there are several other companies looking to build plants in China," the *Crown Gazette* reported that year. "We received verbal orders for up to four extractors and five DTDCs to be confirmed in the next six months!" Joining Crown in these projects was solidly reliable K. K. Yee of Pisces Engineering, who decided to expand his Malaysian fabricating business by supplying China, too.

Whose Process Is It?

The road to these contracts had its potholes. For several years, process designs by leading companies including Crown were widely circulated in China and pirated by local entrepreneurs. "The Chinese government encouraged copying," says Dan Anderson, director of Crown's Asian operations. "Back in the 1980s, most of our equipment sales were to state-owned companies and the Chinese engineers involved with the negotiations did an excellent job of reverse-engineering our systems." To sell its equipment, Crown had to sit with these designers and answer all their questions. The government-owned design institutes compiled the technical information from these meetings and sold Crown's designs to anyone who wanted to buy them. The price was approximately $30,000 and a buyer had about five years to pay for the package. "We knew we were being copied," Anderson says, "when we saw ads in industrial trade magazines and brochures with our equipment. At one point, I remember there were as many as nine companies making our equipment illegally with the full blessing of the Chinese government."

Crown wasn't alone. Innovative companies were having their top products replicated in "knock-off" forms repeatedly—witness bogus 3M™ Post-It® notes flooding the Chinese market.

"For a while we wrote off China," says Anderson. "We found we were beating our heads against the wall. The locals were especially good at copying small capacity plants, where factors like product quality, hexane and steam consumption were not big considerations. As international companies such as Cargill and ADM went to China and brought our expertise, the local companies found they couldn't compete anymore. They saw the success of the internationals and started moving toward our technology and—although they didn't want to pay the price—the successful companies in China started purchasing more and more of our plants and equipment."

By 2000, Crown had decided to have a stronger presence in China, not only to serve its long-time U.S. customers, but also to attract local Chinese companies.

About that time, George Anderson took a call from Maurice Bell in Toronto. "George," Bell said, "I have someone you ought to meet. He's thinking about selling his company and he's a nice guy, an honest guy. He thinks like an American. He's in my office right now."

(Above, Right): Artist's rendering of a Crown Friendship Engineering plant near Wuhan, China. (Left): Crown's partners in China run a world-class manufacturing operation. Pictured (from left) are Luo Xue, founder of Friendship Engineering Company (FEC); Crown's Ralph Romano; Ian Marshall of Crown Friendship Engineering Company; Dan Anderson, director of Crown's Asian operations; and Luo Jun, owner of FEC and Luo Xue's son.

His name was Luo Jun, owner of Friendship Engineering Company (FEC) of Wuhan, China, an industrial city near the famous Yangtze River Three Gorges project. FEC, founded in 1993, had become a leader in supplying low-cost, turnkey extraction and refining projects. Founded by Luo Jun's father, Mr. Luo Xue, FEC's strength was turnkey installations completed in as little as 250 days from receiving the order to final plant commissioning. Crown valued the family's strong technical skills and ethical approach to business. Luo Jun's mother, Madame Xu, was the company's chief engineer.

Maurice Bell hoped to introduce FEC's expertise in America. Lou Jun had considered other buyers, Bell told Anderson, but Crown's culture was a better fit. "Had Maurice not put us together, FEC would have been sold to someone else," says George Anderson. Crown formed a joint venture with FEC in the fall of 2001 and named it Crown Friendship Engineering Company (CFEC).

Nothing Ventured

In another nod to partnerships, Crown crafted an alliance with the oils and fats group of Alfa Laval Oil and Protein Technology in 2000. Part of a large Swedish company founded the same year as Crown, this organization offered expertise in oils and fats refining, including separation equipment, heat exchangers, centrifuges, fluid handling equipment and deodorizers, giving Crown more clout in a relatively new arena. "Alfa's knowledge on many of the refining processes, their existing customer base, proliferation of overseas offices and local procurement expertise has obvious benefits to Crown," Stuart Smith, of Europa Crown, said. "Together, Alfa Laval and Crown can cover all products from oilseed intake and preparation, through extraction to complete refinery solutions. We'll be able to collaborate across a much wider customer base where, in the past, we may well have been in direct competition."

Jesse See, Jesse Do

> "Gov. Jesse Ventura is packing his bags and getting briefed this week for a one-week expedition to China that has been described by his aides as 'the mother of all trade missions.'"
>
> **Star Tribune, June 2002**

Among the nearly 60 Minnesota businesses joining Minnesota's wrestler-turned-Governor on a trade mission to China in 2002 was Crown.

"The company that produced those red Tinker Toy-like things that help hold up the Metrodome is now focused primarily on making machinery that separates edible oil from soybeans, sunflower seeds and other oil-producing seeds," the Star Tribune reported. "Crown has already built a half-dozen oilseed processing plants in China."

The trade mission was reportedly the largest of its kind ever organized by a state government. Ventura, Minnesota's self-appointed "premier salesman and promoter," set the pace with pre-trip reading assignments and a nine-day journey that taxed even the most stalwart travelers. Ventura was convinced that China, which would become the world's second-largest economy, held great promise for Minnesota, especially in agriculture, processed foods, information technology and medical products. From 1999 to 2002, Minnesota's exports to China had grown by 84 percent—more than four times the national rate.

"You naturally want to do business and hang with people that are successful," Ventura told the Star Tribune in his classic colloquial style. "China's the new horizon, the new economic opening out there and we certainly want Minnesota to be in on the ground floor."

Joining George Anderson on the China mission was Bob Carlson, Crown's manager of Crown Friendship Engineering Company (CFEC), and his wife, Mary Jane, residents of Beijing. "Never a dull moment," Anderson reported of the mission. "The Governor is an entertainer and he has celebrity status, plus he is huge, has a shaved head and blushes often when being flattered by the Chinese. This happened at almost every public event."

Anderson said the meetings, site visits and educational presentations were excellent and U.S. trade officials "pulled no punches."

"One statement in our delegation materials said this," Anderson recalled. "The Chinese believe that Westerners are friendly, honest and trusting and that they conduct business with a high level of integrity. However, they don't feel compelled to behave in a similar fashion. Deception and the exploitation of weakness are time-honored strategies in China's business arena."

But Anderson could not fully agree: "This description surely does not fit all Chinese businessmen," he said. "And there are clear signs of good will and many efforts by the Chinese to accommodate American-style business." He cited a U.S. trade official who said he could accomplish more in a year or two in China than during his entire career in Europe.

Accommodating, indeed. To ensure that Minnesota's size 4X Governor would fit a special shop coat for a visit to a Hormel Foods packing plant near Beijing, they fashioned five immense coats. Even on Anderson, exceeding six feet tall, the garment reached nearly to his toes.

Crown's George Anderson joined Minnesota Governor Jesse Ventura and nearly 60 other companies on a trade mission to China in 2002. Their accommodating hosts prepared for their size 4X visitor (the governor) and some of his tall friends.

The advantages of this new alliance were quickly evident, Stuart Smith said. Within a few months of the agreement, a Russian customer who had just seen the successful start-up of a new Crown extraction plant hired Crown again to oversee installation of a new deodorizer supplied by Alfa Laval.

In Crown's time-honored tradition of strengthening relationships over fishing at a north-woods lodge in Minnesota, representatives of Crown, Europa Crown and Alfa Laval met at Ruttger's Bay Lake Lodge in June 2000 for a weeklong technical seminar to learn about each other's product lines and explore how best to work together. They came from England, Mexico, Honduras, Brazil, Argentina, Indiana and Minnesota. And during the free hours, Minnesotans Chas Teeter (a recent transplant from Texas) and Jeff Kraker shared their secrets of walleye fishing.

But that carefully planned partnership was brief, proving the delicate nature of alliances. Not long after it was created, Alfa Laval's management team left the company. They had not had time to bring their large, worldwide organization to believe in a partnership with such a small, non-European upstart as Crown Iron.

The Road to Refining

Ken Carlson came to Crown with Alfa Laval ties. After graduating as a mechanical engineer in Stockholm, Sweden, he worked 14 years with the company and three years with Wurster & Sanger before Crown acquired them. Carlson operated his own consulting firm—specializing in oils and fats refining, bleaching and deodorizing—for 13 years. He consulted with Crown through the 1990s and joined the company in early 2003.

When he worked with Wurster & Sanger, Carlson redesigned their refining technology. "Most important for Crown, they had glycerin technology that produced oleochemicals, the ingredients used to make soap, cosmetics, detergents," says Carlson, "and ultimately biodiesel."

As Technical Director of Oils and Fats, Carlson's role focuses on boosting the company's presence in refining. "Crown and extraction are synonymous," Carlson says, "and now we're moving into refining." Crown is the "new kid" on this block. "Our market presence is still very small compared to our competitors'. Our potential is high because refining is almost as big as extraction all over the world. It's a steady, high-volume business."

How will Crown build that market presence? Knocking on doors, telling the Crown story, building a high-quality, refining "reference plant" with all the bells and whistles, open for all to see.

To give Crown customers a chance to test their refining ideas in the lab, Carlson created what he nicknamed the "Plug and Play" R&D Refinery, a skid-mounted pilot plant expressly for research and customer use. Within weeks of making it operational, customers were putting their names on a waiting list for time in the Crown lab.

A Reluctant Farewell

Although Crown set records for sales and profits through most of the 1990s, it was clear that one division—CAM Manufacturing, Inc.—continued to struggle year after year, with only brief episodes of profitability. "CAM didn't fit Crown's plans for the future and there are

> "We are seeing fundamental changes in the way we do business. Partnerships and alliances are the name of the game and their impact on Crown will be significant."
>
> **Stuart Smith, Europa Crown Limited**

> "This past year has been difficult for all CAM employees. Their work future was in doubt . . . Several left for more secure opportunities, only to return. That was a wonderful compliment to CAM and Crown."
>
> *Crown Gazette,* **Spring/Summer 2004 issue**

(Above, Left to Right): Crown's Allen Ost is dwarfed by a 7,000 tons-per-day Crown extractor in Argentina. A Crown 650 tons-per-day DiFlow Deodorizer installed in a refining plant owned by Riceland Foods in Arkansas.

> "Busy, busy, busy is the quickest way of describing our present activity levels."
>
> **Phil Blenkiron, managing director, Europa Crown Limited**

some interested possible buyers," Cliff Anderson told employees. Crown decided to sell CAM Manufacturing, Inc., in 2004.

Crown employees reminisced about CAM's move from Minneapolis to Cokato, Minnesota, in 1985 and the start of a lunchtime softball team. They recalled an invention called the "Trailer Anchor," designed by Jim Fischer using CAM's venerable earth auger design. They remembered Dean Ailie's commendation for helping save a young woman's life in 1999 and the example he set as a hard-working employee who cared deeply about quality. They relived their history: new employees, retirements, new babies and grand accomplishments of the plucky little venture in Cokato.

Pengo Corporation of Laurens, Iowa, bought CAM Manufacturing with the mutual assurances that CAM would remain a valued Crown supplier. They continue to supply Crown today.

"Busy, Busy, Busy"

Reports from the field in 2004, 2005, 2006 and 2007 were peppered with active verbs and adjectives. Europa Crown's Ken Bell and Clive Musson were "logging up the air miles chasing future contracts with promising project potential."

With increasing demand for oil processing equipment in India and Africa, Crown announced a new joint venture with Kumar Metal Industries, a company based in Mumbai, India, and founded by Onkarnath R. Manaktala in 1939.

Richard Holland and his team arrived in Aston, Russia, to commission a new project, and new orders for Cargill in Russia and Bunge in Poland soon followed.

Noel Rosenthal and Ben Floan started up a 3,000 metric-tons-per-day hot dehulling and

extraction plant for ADM in Brazil *plus* expansion of another ADM plant to a whopping 9,300 metric-tons-per-day in Argentina.

Hernán Paredes and Ryan Popinga started up a new deodorization plant for Industrial Patrona in Cordoba, Veracruz, Mexico, in a record three days.

Alejandro Citterio, Crown's business development manager, was all over Argentina tending to multiple preparation and extraction projects in a country once dominated by Crown's competitors.

Stateside, Crown sold one of the world's largest semi-continuous deodorizers to Riceland Foods, using the trademarked Diflow technology. AVOCA (formerly RJ Reynolds) bought Crown's largest Model IV Extractor to date for use in the nutraceuticals market. And for two years running—2004 and 2005—Crown was named one of Minnesota's top 50 "fastest growing private companies" by the *Minneapolis/St. Paul Business Journal*.

Crown's newly renovated lab in Roseville announced dramatic research and development breakthroughs.

The "Pack Rats" Prevail

Though Crown's commitment to research and development dates back decades to its Northeast Minneapolis Tyler Street plant, the company's new solvent extraction lab, created in 2003, was a massive improvement over any previous Crown facility. The variety of services and confidentiality it offered was clearly a first in the industry.

(Above, Left to Right): Vegetable oil used in kitchens around the world is an end-product of Crown's oleochemical processing work. Precision welding is crucial in producing this leak-proof heat exchanger coil assembly for a Crown deodorizer, a mechanism used in oils and fats refining.

As Published In
THE BUSINESS JOURNAL
SERVING MINNEAPOLIS-ST. PAUL
OCTOBER 2004

50
FASTEST-GROWING
PRIVATE
COMPANIES
2004

Crown Iron Works Company
50 Fastest-Growing Private Companies
The Business Journal – October 2004

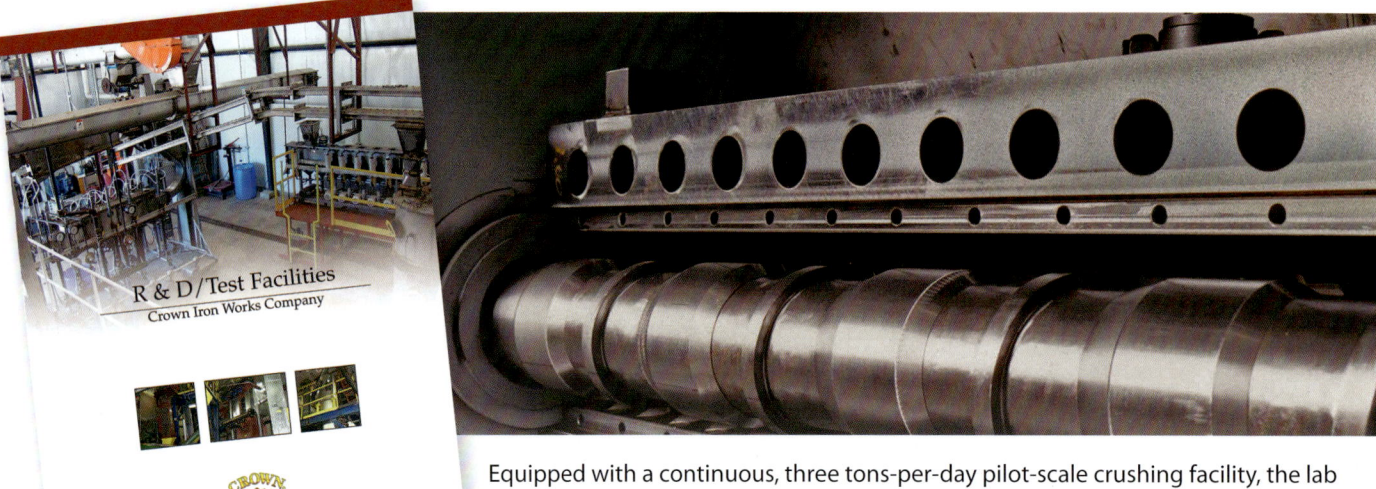

R & D/Test Facilities
Crown Iron Works Company

A press at Crown's Research and Development Lab used in developing HIPLEX® technology (a joint venture company with Harburg Freudenberger of Germany).

Equipped with a continuous, three tons-per-day pilot-scale crushing facility, the lab was designed to develop new applications and test proprietary processes for Crown customers and prospects. More than a third of its work generates revenue.

Chas Teeter, a graduate from research powerhouse Texas A&M University, was given the chance to build the new lab. "Back in 2000, it was three garage stalls with dusty equipment," Teeter recalls. "Grant Angrimson, a lab technician, and I started pulling things out of storage and going to auctions to buy testing equipment. Much to the accounting department's frustration," Teeter says, "we'd buy about $1,500 worth of surplus equipment every month. With that we built the analytical lab and, with equipment Crown already owned (some dating back 30 years), we began assembling the processing lab." They found some great deals ($150 for a fume hood that would cost about $6,000 new) and they managed some useful swaps (a used drill press for a roto vat). Crown people started calling them "the pack rats."

After about seven months, the lab was ready. Its purpose was to support Crown's existing plants and conduct pre-engineering work. At the end of every research project, the lab turned its data over to engineering. In that way, Crown had a process guarantee, ensuring that the equipment tested in the lab would function the same way in the field.

Crown customers or potential customers have access to the Crown lab, its equipment and staff to run tests and advance proprietary research. About 60 percent are Crown's current customers and the other 40 percent are new. They hear about the lab and its capabilities by word of mouth, from trade shows, trade journals or on the Internet. Though most of this work is proprietary, a few examples illustrate the unusual variety of research requests. "In one case Crown was asked to develop a process to extract carotenoid from sea cucumbers," says Teeter. "It's used in some medications, as well as make-up. In another, we explored extracting a component called 'neem' from an African leaf that serves as a natural insecticide."

Not every project leads to a new process or application, but some do. The "germination period" from R&D to engineering, sales and ultimately a new plant and application is often six years.

One of those new, exciting applications involves using CO_2 instead of traditional solvents to extract oil from oilseeds.

When Last Seen...

Jim Kenney knows freight. He's been handling purchasing at Crown for nearly 40 years and import and export shipments for more than half that time. With the company's expansive international growth, those logistics have amplified and multiplied.

Take the shipment of critical processing equipment from the Pisces Engineering fabricating shop in Malaysia to a new plant site in Bolivia. When the equipment was ready and a ship became available, it was rerouted unexpectedly through the Suez Canal, arriving in Germany. Then, it was transferred to another ship crossing the Atlantic and traveling through the Panama Canal down to Chile, where it was unloaded and carried through the mountains to Bolivia. "That was one of Crown's first, big international projects and a lot was riding on it," says Kenney. "We were coming in early making calls to Europe, trying to get the equipment unloaded and put on a different vessel. We tried everything."

The rerouting had delayed delivery by about 45 days, nearly tripling the shipping costs. "It all ended up working out okay," he says, "but the delays and panic were tough and costly."

That project had other extenuating circumstances, including the accidental death of a key Brazilian advisor on a treacherous mountain road and faulty equipment from an untried supplier. In classic Crown style, the company stayed the course to make the project right in the end. "We didn't make any money on that deal," says Cliff Anderson, "but we got through it."

Kenney remembers another panic when Crown's equipment for a turnkey soybean and sunflower seed processing plant, including a refinery and bottling plant, was stranded in Damietta, Egypt, on the Nile River—not months, but years. The customer paid for the equipment with USAID funds, but failed to live up to his responsibilities for the project, namely building a structural foundation for the plant.

After Crown's equipment arrived at the Egyptian port, "We had to unload," Kenney says, "so we purchased close to 50 huge containers to store the equipment on land. It wasn't just one shipment; it was a group of shipments from all over the world." Ultimately, the original owner ran out of money for the project. Seven long years later, the project was resurrected when a new buyer stepped in and moved only the extraction plant installation to Tunisia. Crown's Gary Koerbitz successfully guided the project to completion in 2007. "I think Crown followed the right track, as painful and longwinded as it was," says Cliff Anderson. "Our philosophy is to follow through with things. Our solutions aren't always perfect, but we do what we say we're going to do."

With time, the scale of Crown equipment that travels the globe has grown considerably. "The biggest single thing we had to transport was a deodorizer fabricated in Malaysia and shipped to Stuttgart, Arkansas," says Kenney. "It was about 18 feet in diameter and 120 feet long, weighing 50 tons."

It didn't come apart in pieces.

"We shipped it from Malaysia to New Orleans, then arranged with a barge to meet the vessel in port," Kenney says. "I hired riggers and a trucking company to unload it onto the barge." The truck was super-sized and outfitted for just such a massive load. The barge was then sent up the Mississippi River to the Arkansas River and finally, to the smaller White River. At that point, says Kenney, the water level was too low and the tug pushing the barge hit bottom.

The trucking company located another site on the river that had sufficient water depth; they built a special off ramp for unloading and finished the job.

A huge truck with an especially long trailer to accommodate the deodorizer traveled the final ten miles to the job site.

These transport projects are increasingly complicated, as ever-larger equipment traverses seas, oceans and rivers, says Kenney, and every one is different, requiring ingenuity and endurance. "That's why," he says, "I haven't got much hair left."

A Day in the Life of Sam Soybean Today

Today, Crown applies its engineering expertise to all aspects of oilseed processing, including preparation, extraction, degumming/neutralizing, bleaching and deodorizing.

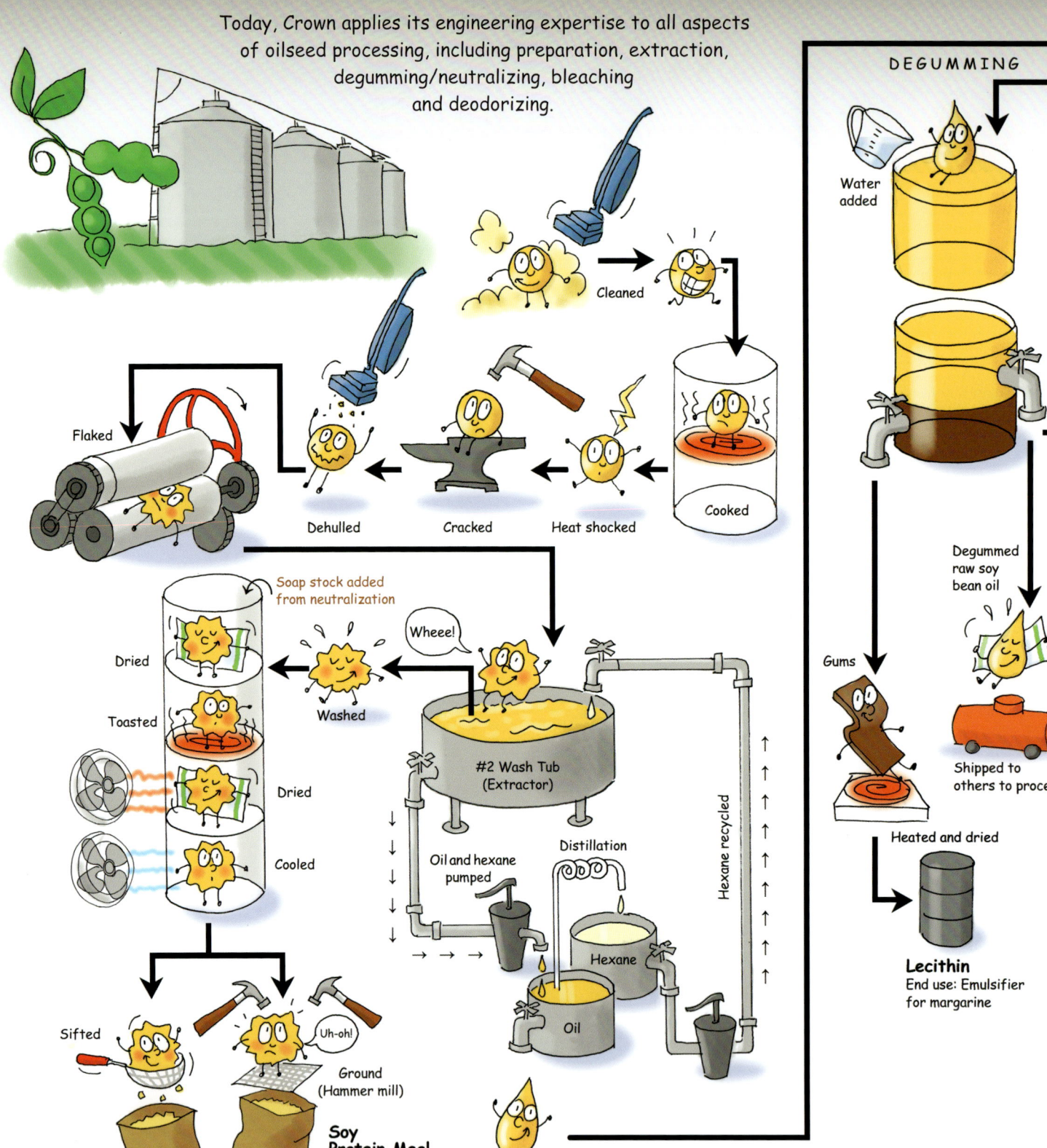

DEGUMMING

Water added

Cleaned

Flaked

Dehulled

Cracked

Heat shocked

Cooked

Degummed raw soy bean oil

Gums

Shipped to others to process

Heated and dried

Lecithin
End use: Emulsifier for margarine

Soap stock added from neutralization

Dried

Toasted

Dried

Cooled

Washed

Wheee!

#2 Wash Tub (Extractor)

Oil and hexane pumped

Distillation

Hexane

Oil

Hexane recycled

Sifted

Ground (Hammer mill)

Uh-oh!

Soy Protein Meal
End use: Food for people and animals

Crude soy bean oil

*Sam Soybean was created by Quincy Soybean Company in 1962.

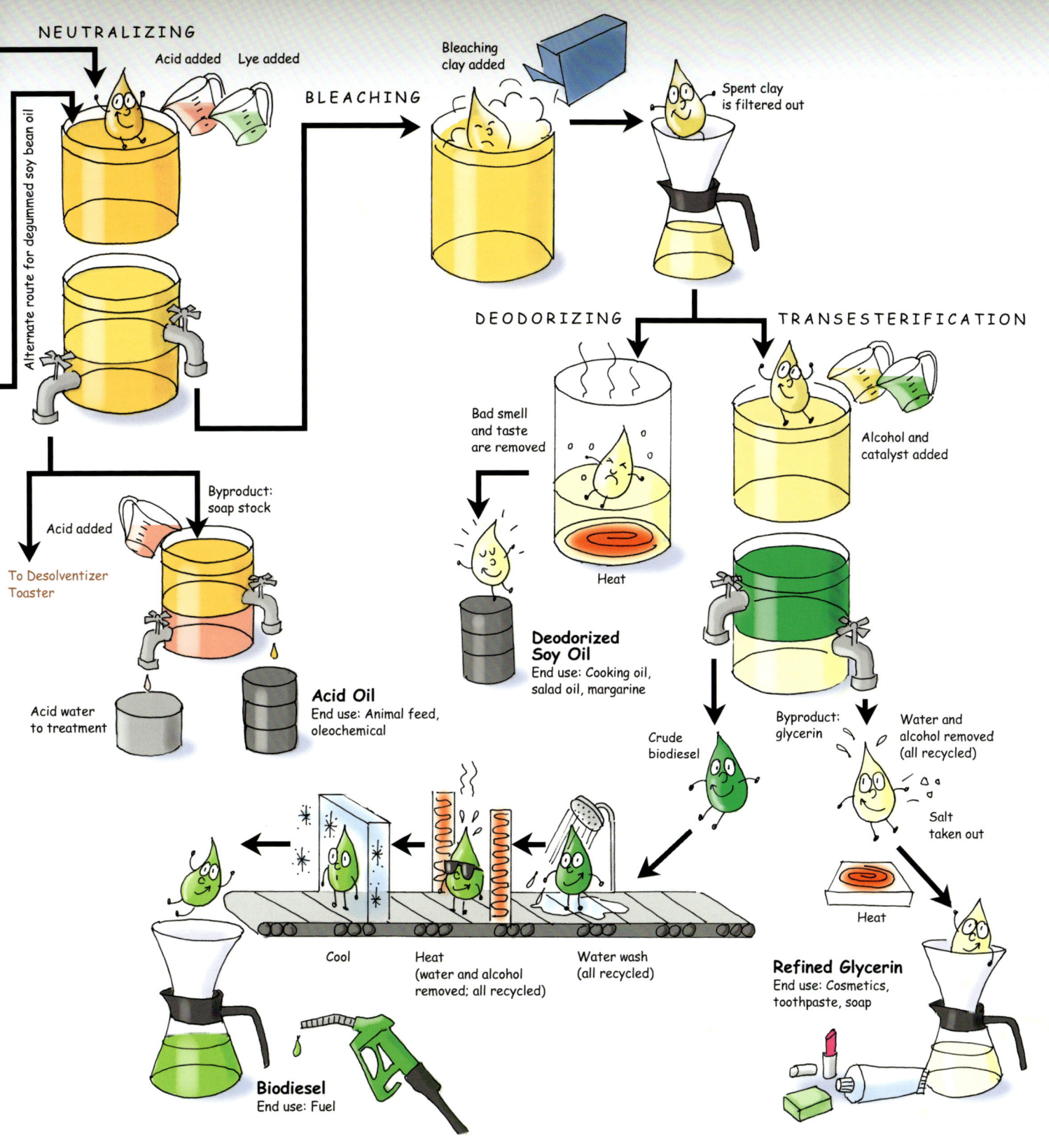

NEUTRALIZING

Acid added   Lye added

Alternate route for degummed soy bean oil

BLEACHING

Bleaching clay added

Spent clay is filtered out

DEODORIZING   TRANSESTERIFICATION

Bad smell and taste are removed

Alcohol and catalyst added

Heat

Acid added

Byproduct: soap stock

To Desolventizer Toaster

Acid water to treatment

Acid Oil
End use: Animal feed, oleochemical

Deodorized Soy Oil
End use: Cooking oil, salad oil, margarine

Crude biodiesel

Byproduct: glycerin

Water and alcohol removed (all recycled)

Salt taken out

Heat

Cool   Heat (water and alcohol removed; all recycled)   Water wash (all recycled)

Refined Glycerin
End use: Cosmetics, toothpaste, soap

Biodiesel
End use: Fuel

(Above, Left to Right): Crown's trademarked HIPLEX® process uses CO_2 injected at high pressure to yield more oil from oilseeds. Nikolaus Foidl, an inventor based in Nicaragua, holds the patent for injecting fluids into mechanical presses. He approached Crown with his idea and helped make this new technology possible.

> "At first, CO_2 looked like it had the potential to replace solvent extraction. If there was going to be a replacement, Crown wanted to be the one to find it."
>
> **Bruce MacKinnon, Crown's product sales manager for CO_2 presses**

Into the CO₂LD®

Bruce MacKinnon comes from a long line of engineers: his father, brother, sister, multiple uncles and his grandfather are all engineers. MacKinnon earned his chemical engineering degree from Michigan Tech and joined Crown in 2002. Specializing in process control, MacKinnon witnessed Crown's early research in supercritical CO_2, a fluid that enhances mechanical oil extraction.

"An Austrian inventor named Nikolaus Foidl, living and working in Nicaragua, had approached Crown around 1999 about injecting alcohol in the mechanical pressing process to extract oil," says MacKinnon.

"Nikolaus held the patent for injecting fluids into mechanical presses," Jeff Scott remembers. "He went to all the major screw press manufacturers with his idea—Krupp, French Oil Mill, Rosedowns—and they said they weren't interested. Then Nikolaus came to us. We were the only company that wasn't really a manufacturer of screw presses." Acting on Crown's philosophy of always keeping the door open to new ideas, Scott took a good look at Foidl's idea.

There was no question that Foidl's work helped advance Crown's research, but CO_2 seemed the best bet. Crown's idea was to inject CO_2 at 10,000 pounds per square inch in a "closed cage" environment advocated by Foidl, but problems surfaced. Then Crown tried an open cage in a standard press, with more success. The process used considerably less CO_2 and yielded more oil from the seeds. It led to a patented process and registered trade name that Crown calls HIPLEX®.

In a standard mechanical press, the process might leave 8 percent of the oil in the meal, Mackinnon says, but with CO_2 extraction, that percentage can be reduced to as low as 3 percent. Almost immediately, Crown realized that this discovery could be used on corn germ, a component of corn left over from the ethanol process.

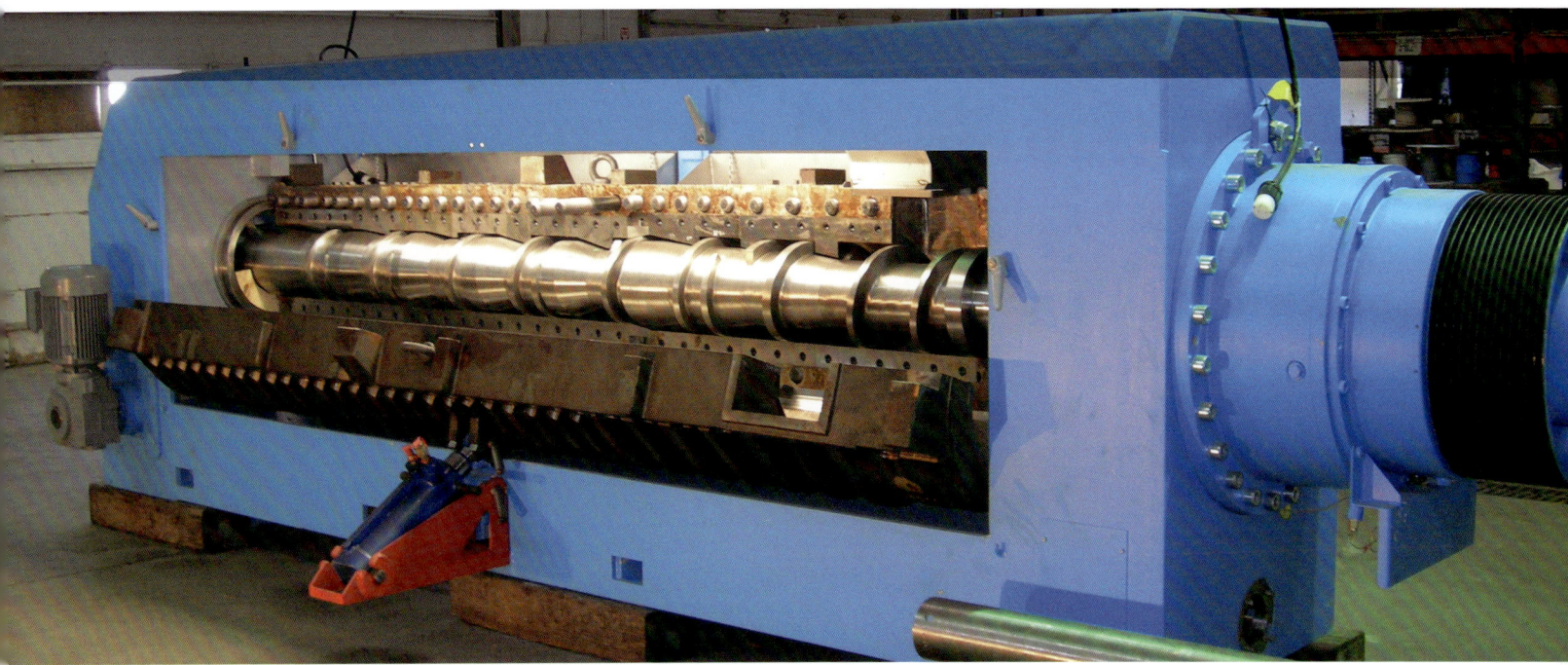

CO$_2$ extraction isn't a new idea. Crown engineers remember conversations about it years ago at ADM. European engineers had been exploring the idea, too. In fact, among those researchers were engineers Jens Schulz and Günter Simon of Harburg Freudenberger, Hamburg, Germany, once a small division of Krupp. "We had the best screw press technology and we believed that CO$_2$ would make the de-oiling process better," says Schulz, director of engineering for Harburg Freudenberger's edible oils division. "From a mechanical point of view, there are limits to how much oil can be extracted. CO$_2$ promised far better results, almost on a par with hexane extraction."

At an American Oil Chemists' Society (AOCS) conference in America, Schulz and Simon approached Jeff Scott. The prospect of an alliance was palpable. "Our core business is screw presses," says Simon, "and Crown's core business is extraction. Our main market is Europe and Crown's is the United States and the Americas. We believed we could cooperate and that's why we approached them. We were frankly surprised by their openness."

Crown and Harburg Freudenberger formed an alliance in early 2007 called HIPLEX®, a technology holding company and new business with 50/50 ownership. Their focus is using CO$_2$ technology to achieve a higher oil recovery rate than conventional pressing to produce high-quality animal feeds. And—with nearly 200 existing and projected ethanol and biodiesel plants in the United States—the new process has compelling value.

Another business, called CO$_2$LD®, will use the new technology to produce high-protein meal for human consumption. By controlling temperatures inside the mechanical press and maintaining a cool environment, the protein retains its highest nutritional value. In a world where organic, safe, vitamin-packed foods are popular, CO$_2$LD® has countless applications.

(Above): This Harburg-Freudenberger/HIPLEX® Press, modified by Crown, uses CO$_2$ technology to achieve a higher oil recovery rate than conventional mechanical pressing. The end-result is CO$_2$LD® high-protein meal for human consumption, a popular ingredient in organic, safe, vitamin-packed foods.

"Crown's Ben Floan discovered one of the key breakthroughs when he found a reliable method for injecting CO$_2$ into the machines. He has a patent with his name on it for the injector."
Jeff Scott

"Anybody doing
mechanical pressing
can get more oil
out of the seed with
supercritical CO₂. It's a
perfect application for
plants producing below
500 tons-per-day."

Bruce MacKinnon

Locally Grown
Food *and* Fuel

*Raised on a farm in Ellsworth, Iowa, and trained in business, Mark Hill was looking for
a new enterprise in 2003. After learning about agribusiness in the seed industry and
becoming the largest independent soybean seller for AgriPro, he was restless.*

*"I wanted to move further upstream in the food industry," he says. "I started working
with Dr. Larry Johnson at Iowa State University, to explore how to capture oil from
soybeans using a more natural process. I wanted to appeal to the growing natural foods
business in America."*

*Johnson put Hill in touch with Jeff Scott at Crown. "Though the original process
was pretty primitive in their lab," he says, "I saw the high-quality pressed oil they were
producing. I was amazed at the potential."*

*Hill decided to hitch his business plan to Crown and CO_2LD®. He founded Safesoy
with a focus on producing high-quality protein and oil for the natural foods market.
Later, he expanded his sights to include high-quality animal feed and applying the new
CO_2LD® technology to renewable fuels.*

*Hill's plant in Ellsworth, Iowa—started up in 2007—is an R&D facility in many ways.
Safesoy, Crown and Iowa State are collaborating on the project, which, if successful,
could revolutionize the oilseed industry and give small, independent farmers new
opportunities. "This processing system and plant configuration would be purchased by
local producer groups," says Hill. "It can be a community asset that not only produces
high-quality food and fuel, but it creates jobs and generates revenues for local
communities."*

When Crown Companies celebrated its 130th anniversary in 2008, Cliff Anderson, the company's president for 34 years, retired.

What Makes Us Proud?

By the time Crown Companies reached its 130th year in 2008, it had much to be proud of—persistence in the face of calamity, creativity and reinvention in volatile and changing markets, determination to deliver the highest quality no matter what. A desire to leave a legacy that values people, honesty and reliability.

Shinji Ikemori, a Crown partner at Techno Sigma, Inc., in Japan is proud of Crown's "good reputation, excellent technology and the high esteem Crown receives from its customers."

Hershel Austin, retired from Cargill, says Crown has reason to be proud of its commitment to the fundamentals: "Crown has always been there to satisfy the customer," he says. "Their philosophy has been, 'keep it simple, safe and efficient.' If there was a problem, we could always sit down, talk and make it right."

"Over my 40 years, I have seen so many of us 'in our element' with excellent cooperation and mutual support. Each person seems to feel a similar pride and enthusiasm for their work."

George Anderson

> "I'm proud of the progress Crown has made over all these years. The commitment they've demonstrated and the loyalty they've inspired is rare in business today."
>
> **Leroy Venne, former employee and Crown customer**

Stan Loft, a Crown competitor for many years, appreciates the uniqueness of Crown's approach to oilseed extraction: "When DeSmet came to the U.S. in the 1970s, they said they had a duplicate of Crown's extractor," he says, "but they were never any competition for Crown. Their extractor was truly unique." Loft knows that top performance comes from excellent people. "With any engineering company, it's the people who make it," says Loft, "and Crown takes care of its people. If I was starting an engineering company, I'd try to hire some of them away. They're that good."

Gary Pulis, an engineer with Crown since 1978, is proud of "our freedom to work on a moral basis. We once had contracts for a million dollars that were one-page in length. People can trust us . . . and that has always made it easy to come to work."

"Crown retains its people and I appreciate that," says Leo Gingras, vice president of billon-dollar Riceland Foods. "They can offer continuity on projects and that's unusual in this era. Crown people are very personable and professional. It's easy to build a good working relationship with them."

"What makes me most proud are the ethics of our company," says Dan Anderson. "We're a market leader in many of our products and certainly second or third in others. But I think the reputation of Crown is its greatest strength. Cliff and George will lose money on a job as opposed to doing work on the cheap."

Dennis Wendland, senior vice president of oilseed processing at Cenex Harvest States (CHS), calls Crown the "premier company." "Somebody could argue that they're number two in the world, but I think they're premier. That's really saying something for a company from Roseville, Minnesota."

The 2007 Crown team—or at least the ones who weren't away on a start-up somewhere else in the world.

In Closing . . .

As we look to Crown's future, the possibilities are endless.

Ideas and technologies will lead Crown to new markets in new places that we cannot predict—any more than our ancestors could imagine Crown becoming a process technology leader serving customers on all the major continents of the world.

But compared to our early history, the world poses new priorities that Crown will need to address: atmospheric change, pollution, recycling, energy and water shortages, dietary concerns and greater demands on the world's food supply.

We challenge Crown to step up to this changing world, just as it has in the past.

We believe in Crown because one of its key strengths is resilience. It has consistently demonstrated the ability to respond to change with improved capabilities and a positive attitude. Crown will continue to grow toward the future.

Cliff Anderson

George Anderson

About the image above: This Crown logo, rendered in stained glass, was designed by Clifford H. Anderson in the 1930s. The glass came from a window that hung behind Anderson's desk at Crown's Tyler Street plant in Northeast Minneapolis.

Crown Employees (2000–2008*)

Clifford Anderson, 2/65
Dale Stolt, 11/65
John Chasteen, Jr., 11/67
James Kenney, III, 4/68
George Anderson, 6/69
Richard Grabow, 6/70
Dennis Cox, 4/71
Marie Nichols, 9/72
Jeffrey Scott, 10/77
Gary Pulis, 1/78
Sharon Deason, 2/79
Jeffrey Christopherson, 7/79
Paul Ell, 1/80
Sharon Trocke, 8/80
Dean Nordquist, 12/80
Alan Loerzel, 1/81
Susan Young, 8/87
Robert Deleo, 5/88
Christine Johnson, 10/88
Philip Smith, 3/90
Allen Ost, 11/90
Patrick Hudoba, 6/91
Timothy Maneely, 8/91
Gregg Haider, 2/92
Daniel Anderson, 8/92
Grant Angrimson, 9/92
Jeffrey Kraker, 3/93
Corey Paulson, 6/94
Janice Shear, 8/94
Kimberley Robert, 10/94
Katherine Liesmaki, 11/94
Noel Rosenthal, 1/95
William McDonald, 2/95
Hernán Paredes, 3/95
James Sandkamp, Jr., 6/95
Derek Masterson, 9/95
Fritz Beckman, 9/95

Paul Andrews, 2/96
Richard Ozer, 12/96
Oliver Kachkovsky, 8/97
Yu-Hsien Wu, 5/98
Gerald Austin, 9/98
Corrine Sorensen, 5/99
Jeffrey Scott, Jr., 6/99
Mathew Van Someren, 6/99
Ralph Romano, Jr., 8/99
Ann Loveland, 9/99
Kameron Carlson, 11/99
Francesca Welsh, 11/99
Sheila Baker, 12/99
Charlie Teeter, Jr., 1/00
SueEllen Altrichter, 2/00
Alejandro Citterio, 2/00
Kathleen Halverson, 2/00
Gary Terhaar, 2/00
Christina Campbell, 5/00
Rosina Ortwein, 5/00
Lon Smallridge, 9/00
Gary Koerbitz, 9/00
Ryan Popinga, 10/00
Johann Hasslbeck, 11/00
Kristopher Colbert, 1/01
Erik Monson, 5/01
Rigoberto Merino, 6/01
Warren Kroeker, 6/01
Heather Johnson, 8/01
Richard Watson, 9/01
Steven Rogers, 10/01
John Soukup, 10/01
Raymond Johnson, 10/01
Wayne McCalley, 2/02
Suzanne Michaud, 2/02
Donatella Baglo, 5/02
Stephanie Bishman, 5/02

Andrew Burgess, 5/02
Dina Courneya, 8/02
Serken Akben, 8/02
Lawrence Sullivan, 8/02
Daniel Wareham, 10/02
Bruce Mackinnon, 11/02
James Benson, 1/03
Nicole Trocke, 3/03
Louise Thurler, 4/03
Benjamin Floan, 5/03
Patricia Japs, 5/03
Angela Liesmaki, 5/03
Gerald McBurney, 6/03
Michele Liesmaki, 6/03
Janel Olson, 6/03
Cynthia Trocke, 6/03
Michael Rannow, 7/03
Kenneth Carlson, 7/03
Curtis Dahl, 8/03
Haskell Cooke, 2/04
Nick Rosenthal, 3/04
Tiki Cartwright, 3/04
Eva Christopherson, 6/04
Rishabh Maniktala, 7/04
Mark Carlton, 8/04
Ellen Handrahan, 9/04
Gregory Waranica, 10/04
Catherine Kallaus, 6/05
Pedro Macedo, 8/05
Frank Walek, 11/05
Alexander Danelich, 2/06
Patrick Harrington, 2/06
David Mannello, 2/06
Philip Fisher, 3/06
Frank Bennett, 5/06
Jeffrey Wheelis, 5/06
Ryan Romano, 5/06

* Those listed in black were current employees as of August 2007.

Wayne Lazerte, 6/06
William Morphew, 8/06
Scott Taff, 8/06
Lee Snoeyenbos, 8/06
Jerome Hawker, 8/06
Jennifer Abara, 9/06
Bruce Cook, 10/06
Robert Jacobson, 10/06
Paulo Telles, 11/06
Sean Mathwig, 11/06
Nicole Hudella, 11/06
Anton Matwiejko, 11/06
Mark Haas, 12/06
Thomas Myers, 12/06
Joshua Vincent, 1/07
Erin Olsen, 1/07
Lisa Reich, 2/07
Eric Karis, 3/07
Karl Englund, 4/07
Jesse Devine, 6/07
Brad Kletzin, 6/07
Katie Liesmaki, 6/07
Arthur Christopherson, 7/07
Madeline Carlson, 9/07
Matthew Olsen, 9/07
Mary Berg, 10/07
Lucas Strassburg, 10/07
Charlene Kelly, 12/07
Tina Schachel, 1/08
Matthew Ducharme, 3/08
Nathan Menges, 4/08
Jeff Slowiak, 5/08
Jim Hoel – Contract
Paul Schiller – Contract
Eugene Knurenko – Contract

Europa Crown Limited

Ken Bell, 1/90
Peter Wilson, 1/90
Lloyd Walker, 3/90
Clive Musson, 5/91
Richard Young, 10/91
Richard Holland, 1/95
Phil Scull, 5/96
Bob Paffley, 2/97
Phil Blenkiron, 10/97
Dave Lockham, 4/98
Linzy Settle, 5/99
Steve Tune, 4/00
Phil Barnes, 8/00
Pam Dunn, 9/01
Andy Nicholson, 9/01
Barry Eagle, 2/02
Malcom Scrivener, 9/02
Lisa Smith, 9/02
Claire Holland, 10/02
John Wilkinson, 9/03
Andy Davis, 10/03
Ian Bithell, 11/03
Helen Hoise, 4/04
Paul Gallagher, 6/04
Kevin Shadlock, 6/04
Clare Harrison, 8/04
Richard Benson, 9/04
Terry Dickinson, 9/04
Paul Hayward, 1/06
Malcom Portor, 1/06
Katie McDonald, 7/06
Liz Read, 7/06
John Rodriguez, 8/06
Bill Tattersfield, 10/06
David Thirsk, 10/06

Liz Sorenson, 3/07
Michelle Larter, 10/07
John Bellenie, 11/07

Valentin Zaletkin, 11/94 (Russia)
Boris Solovyov, 11/01
Julia Ruzhova, 5/05
Eugene Dolgirev, 4/07

Vladimir Porshnev, 7/01 (Ukraine)

Plamen Vasilev, 4/06 (Bulgaria)

Crown Friendship Engineering Company

Mr. Luo Jun, 10/01
Madam Xu Dao Ying, 10/01
Mr. Luo Xue Nian, 10/01
Mr. Luo Wei, 10/01
Mr. Wang Yu Ping, 10/01
Mr. Zhang Jia Xing, 10/01
Mr. Ian Marshall, 3/03

Crown Management Team (August 2007)

(Left to Right): Gary Koerbitz, vice president operations; Clifford I. Anderson, president; George Anderson, vice president engineering; Ralph Romano Jr., vice president finance; and Jeffrey Scott, vice president sales and marketing.

Crown Office Locations

Crown Iron Works, Roseville, Minnesota, USA – Headquarters

Europa Crown Limited, Hessle, England – Headquarters

Rosario, Argentina – Office

San Pedro Sula, Honduras – Office

Ciudad Obregon, Mexico – Office

São Paulo, Brazil – Office

Moscow, Russia – Office

Kiev, Ukraine – Office

Crown Friendship Engineering Company, Beijing, China – Joint Venture

Intecnial, Erechim, Brazil – Joint Venture

Kumar Metals, Mumbai, India – Joint Venture

Crown Trademarks and Brands

CO$_2$LD®

HIPLEX®

DDD®

Hulloosenator®

Tripoint®

Acknowledgments

This book was written by Carol Pine, whose enthusiasm and perseverance carried us through. The team behind her was led by Corrie Sorensen, who contributed long hours and patience in coordinating the project. The team came together to deliver brilliantly, producing a work we are quite proud of. We hope you enjoy this book as much as we have enjoyed making it.

The Crown Team

Carol Pine – author
Corrie Sorensen – editor and project manager
Paul Déak and Cathy Spengler – book designers
Nickie Dillon – proofreader
Terri Hudoba – indexer
Elizabeth Cleveland and Friesens – printer

Editorial Team

Cliff Anderson
George Anderson
Jeff Scott
Carol Pine
Corrie Sorensen

The following people provided their insights and memories for this book:

Cliff Anderson
Dan Anderson
George Anderson
Hershel Austin
Ken Bell
Phil Blenkiron
Frank Boling
Glenn Brueske
Ken Carlson
Jeff Christopherson
Ed Churchill
Alejandro Citterio
Haskell Cooke
Dennis Cox
Alcir Dall'Agnol
Myron Danzer
Sharon Deason
Paul Ell
Jerry Fawbush
Dennis Garceau
Leo Gingras
Joe Givens
Gregg Haider
Blake Hendrix

Marian Herwig Berglund
Mark Hill
Mr. Shinji Ikemori
Dr. Lawrence Johnson
Luo Jun
Marge and Steve Kaiser
Jim Kenney
C. Louis Kingsbaker
Gary Koerbitz
Bill Kratochwill
Stan Loft
Pedro Macedo
Bruce MacKinnon
Sunil Manaktala
Ian Marshall
Derek Masterson
Bill McDonald
N. Hunt Moore
Clive Musson
Allen Ost
Richard Ozer
Hernán Paredes
Corey Paulson
Ryan Popinga

Vladimir Porshnev
Gary Pulis
Ralph Romano
Jeffrey Scott
Jens Shulz
Bill Shurtleff
Günter Simon
Augusto Skyrzpek
Lon Smallridge
Phil Smith
Jeff Stroburg
Chas Teeter
Paulo Telles
Leroy Venne
Lloyd Walker
Greg Waranica
Dennis Wendelen
Norm Witte
Richard Young
Valentin Zaletkin

Photo Credits

Images used in this book were gathered from the Crown archives and from the following sources:

Pages v, 43, 105, 107, 108 and front cover – photographs by Steve Niedorf.

Page 52 – photograph by Dave Santos of H. Larson Photography, Minneapolis, MN.

Pages 98–99 – illustration by Nancy Meyers Illustration, Minneapolis, MN.

Page 120 – photograph from the Minneapolis Public Library collections.

Index

On the following page: Crown circa 1892, flanked by the huge exposition center in downtown Minneapolis, where William McKinley was chosen as the Republican nominee for the presidency.